Target

Get back on track

7

Edexcel GCSE (9-1)
Mathematics
Number and Statistics

Diane Oliver

Pearson

Published by Pearson Education Limited, 80 Strand, London, WC2R ORL.

www.pearsonschoolsandfecolleges.co.uk

Text © Pearson Education Limited 2017
Typeset by Tech-Set Ltd, Gateshead
Original illustrations © Pearson Education Ltd 2017

The right of Diane Oliver to be identified as author of this work has been asserted by her in accordance with the Copyright, Designs and Patents Act 1988.

First published 2017

19 18 17
10 9 8 7 6 5 4 3 2

British Library Cataloguing in Publication Data
A catalogue record for this book is available from the British Library

ISBN 978 0 435 18336 3

Printed in Italy by Lego S.p.A

Helping you to formulate grade predictions, apply interventions and track progress.

Any reference to indicative grades in the Pearson Target Workbooks and Pearson Progression Services is not to be used as an accurate indicator of how a student will be awarded a grade for their GCSE exams.

You have told us that mapping the Steps from the Pearson Progression Maps to indicative grades will make it simpler for you to accumulate the evidence to formulate your own grade predictions, apply any interventions and track student progress.

We're really excited about this work and its potential for helping teachers and students. It is, however, important to understand that this mapping is for guidance only to support teachers' own predictions of progress and is not an accurate predictor of grades.

Our Pearson Progression Scale is criterion referenced. If a student can perform a task or demonstrate a skill, we say they are working at a certain Step according to the criteria. Teachers can mark assessments and issue results with reference to these criteria which do not depend on the wider cohort in any given year. For GCSE exams however, all Awarding Organisations set the grade boundaries with reference to the strength of the cohort in any given year. For more information about how this works please visit: https://qualifications.pearson.com/en/support/support-topics/results-certification/understanding-marks-and-grades.html/Teacher

Each practice question features a Step icon which denotes the level of challenge aligned to the Pearson Progression Map and Scale.

To find out more about the Progression Scale for Maths and to see how it relates to indicative GCSE 9–1 grades go to www.pearsonschools.co.uk/ProgressionServices

Contents

Useful formulae

Unit 1 Statistics

$$\text{frequency density} = \frac{\text{frequency}}{\text{classwidth}}$$

Unit 2 Indices and recurring decimals

$(x^m)^n = x^{m \times n}$

Unit 6 Direct and inverse proportion

Direct proportion: $y = kx$

Indirect proportion: $y = \frac{1}{x}$

Unit 7 Accuracy and bounds

Upper bound A − B = upper bound A − lower bound B

Lower bound A − B = lower bound A − upper bound B

$$\text{upper bound } \frac{A}{B} = \frac{\text{upper bound A}}{\text{lower bound B}}$$

$$\text{lower bound } \frac{A}{B} = \frac{\text{lower bound A}}{\text{upper bound B}}$$

Glossary

Unit 1 Statistics

Sampling: a technique used to gather data. A sample should be representative of the population from which it is taken.

Frequency: how many times something occurs.

Frequency density: calculated by dividing frequency by class width.

Histogram: a way of displaying data. The height of each bar is the frequency density and the area of the bar is the frequency.

Unit 2 Indices and recurring decimals

Recurring decimal: a number where the digits after the decimal point keep repeating forever.

Fractional index: an index that tells you to find a root of the number.

Negative index: an index that tells you to work out the reciprocal of the number.

Unit 3 Surds

Surd: the square root of a number that cannot be simplified.

Rationalising the denominator: removing the surd from the bottom of a fraction.

Unit 4 Cumulative frequency

Cumulative: increasing in quantity as more is added.

Cumulative frequency graph: a graph where cumulative frequency is on the vertical axis.

Box plot: a plot that shows the median, quartiles, and the maximum and minimum values of a data set.

Unit 5 Probability

Independent events: two or more events are independent if the outcome of one does not affect another.

Dependent events: where one event affects the outcome of another.

Conditional probability: the probability of one event happening given that another event has happened.

Tree diagram: a diagram that can help you work out the combined probabilities of more than one event.

Venn diagram: a diagram representing sets of outcomes as circles. Circles overlap to show the common elements of sets.

Unit 6 Direct and inverse proportion

Direct proportion: where one quantity is related to another by an equation of the form $y = kx$

Inverse proportion: where one quantity is related to another by an equation of the form $y = \frac{1}{x}$

Unit 7 Accuracy and bounds

Lower bound: the smallest possible value, given the rounded measurement you are told.

Upper bound: the largest possible value, given the rounded measurement you are told.

① Statistics

This unit will help you to choose and justify the best sampling method to use and to draw and use histograms.

AO1 Fluency check

① Write a sentence to explain each term.

a population ....

b census ....

c sample ....

② Explain a method you could use to take a random sample. ....

....

③ **Number sense**

Work out 10% of

a 250 .... **b** 130 .... **c** 85 .... **d** 124 ....

Key points

| Random and stratified sampling are the two sampling techniques used at GCSE level. | In a histogram the area of each bar represents the frequency. |

These **skills boosts** will help you to identify when to use random or stratified sampling, draw and interpret histograms and estimate the mean from a histogram.

❶ Sampling ❷ Drawing histograms ❸ Estimating the mean from a histogram

You might have already done some work on statistics. Before starting the first skills boost, rate your confidence with these questions.

①
In Maybank High School, there are 100 students in Year 7, 150 in each of Years 8, 9 and 10, and 200 in Year 11. Which is the best sampling method to use to take a sample of 75 students? Explain your answer.

....

②
Work out the frequency density for the class interval $10 < t \leqslant 20$ when the frequency is 53.

....

....

....

③
Work out the frequency for the class interval $85 < h \leqslant 90$ when the frequency density is 1.6.

....

....

....

How confident are you?

1 Sampling

A sample should be representative of the population from which it is taken.
In a random sample every item in the population has the same chance of being chosen.
In a stratified sample the population is divided into groups that have something in common.
The number of items taken from each group is proportional to the size of the group.

Guided practice

Jake records the number of students in each year group in his school.
He wants to take a sample of 80 students to take part in a survey.
Should he take a random sample or a stratified sample?
Explain your answer.

Year group	Number of students
7	200
8	170
9	180
10	130
11	120

Jake could take a random sample by giving all 800 students a number and choosing 80 numbers at random.
Sample size = 80
Total number of students = 800

$80 \div 800 \times 100 = $ ....%
The best sampling method would be ....

because ....

Everyone has an equal chance of being chosen.

Jake could take a stratified sample by choosing ☐% of the students in each year group.
The number of students in the sample from each year group will be proportional to the number of students in the year group.
This makes the sample more representative of the population.

(1) A bank manager wants to survey a sample of 10% of its customers.
Should she take a random sample or a stratified sample?
Explain your answer.

....

(2) The manager of a leisure centre wants to survey a sample of 50 members.
The table shows the ages of the members.
Should he take a random sample or a stratified sample?
Explain your answer.

....

....

Age	Frequency
$16 < a \leqslant 25$	97
$26 < a \leqslant 35$	124
$36 < a \leqslant 45$	102
$46 < a \leqslant 55$	86
$56 < a \leqslant 85$	71

Exam-style question

(3) **a** Describe a method for taking a random sample. ....

.... (1 mark)

b State an advantage of taking a random sample. .... (1 mark)

Reflect A school has 100 students in each of the five year groups. Jai takes a random sample of 50 students. Will there be 10 students from each year group in his sample?

2 Drawing histograms

The height of each bar on a histogram is the frequency density.
Frequency density = frequency ÷ class width

Guided practice

The table shows the ages of the residents in a village.

Age	Frequency
0 < a ≤ 20	456
20 < a ≤ 40	1123
40 < a ≤ 60	828
60 < a ≤ 80	332
80 < a ≤ 100	87

Draw a histogram to show the ages of the residents of the village.

Extend the table with a column for class width and a column for frequency density.

Age	Frequency	Class width	Frequency density
0 < a ≤ 20	456	20	456 ÷ 20 = 22.8
20 < a ≤ 40	1123		
40 < a ≤ 60	828		
60 < a ≤ 80	332		
80 < a ≤ 100	87		

Class width = 20 − 0 = 20,
40 − 20, etc.
Frequency density
    = frequency ÷ class width

Ages of village residents

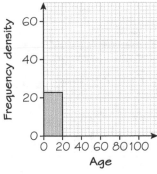

① Ebele measures the heights of some Year 11 students.
The table shows her results.
Draw a histogram for Ebele's data.

Height, h (cm)	Frequency		
160 < h ≤ 165	2		
165 < h ≤ 170	8		
170 < h ≤ 175	11		
175 < h ≤ 180	6		
180 < h ≤ 185	5		

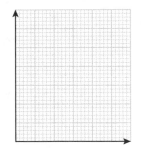

2 The table shows some information about the times, in minutes, taken by a group of students to travel to school on Monday.

Time, t (minutes)	Frequency		
$0 < t \leqslant 5$	20		
$5 < t \leqslant 10$	30		
$10 < t \leqslant 15$	55		
$15 < t \leqslant 20$	60		
$20 < t \leqslant 25$	5		

Draw a histogram for the information in the table.

3 Jamie works in a customer helpline department.
The histogram shows information about the lengths of time, t minutes, of the phone calls Jamie received in one week.
Use the histogram to complete the frequency table.

Time, t (minutes)	Frequency
$0 < t \leqslant 5$	
$5 < t \leqslant 10$	
$10 < t \leqslant 15$	
$15 < t \leqslant 20$	
$20 < t \leqslant 25$	

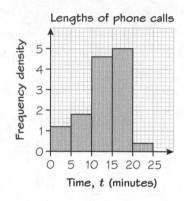

Lengths of phone calls

Hint

Frequency ⟶ ÷ class width ⟩ ⟶ Frequency density

Frequency ⟵ ⟨ × class width ⟵ Frequency density

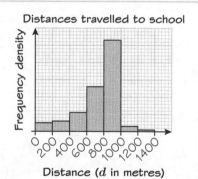

Distances travelled to school

4 The histogram shows some information about the distances, in metres, a group of students travel to school.
25 students travel 200 metres or less to get to school.

a Write down the modal class for the distance travelled to school.

.... (1 mark)

b Work out how many students travelled more than one kilometre to get to school.

.... (2 marks)

c Work out how many students were included in this survey.

.... (2 marks)

Reflect What does the area of each bar represent?

3 Estimating the mean from a histogram

When data is displayed in a histogram, first draw a grouped frequency table and then use it to estimate the mean.

Guided practice

Worked exam question

The histogram shows some information about the heights, in metres, of a sample of trees in a wood.

Estimate the mean height of the trees.

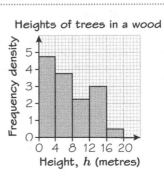

Heights of trees in a wood

Use the histogram to draw a grouped frequency table.

Height, h (metres)	Frequency*	Midpoint	Frequency × Midpoint
$0 < h \leqslant 4$	19	2	$19 \times 2 = 38$
$4 < h \leqslant 8$			
$8 < h \leqslant 12$			
$12 < h \leqslant 16$			
$16 < h \leqslant 20$			
Total			

*Frequency = frequency density × class width

Estimated total height of all the trees = ...

Total number of trees = ...

Estimate of mean = ... ÷ .... = ...

Estimate of the mean = estimated total height of all the trees ÷ total number of trees

① The histogram shows information about train delays, d, from a station during June.

Use the histogram to complete the frequency table and then estimate the mean length of delay.

Delay, d (minutes)	Frequency		
$0 < d \leqslant 20$			
$20 < d \leqslant 40$			
$40 < d \leqslant 60$			
$60 < d \leqslant 80$			
$80 < d \leqslant 100$			
$100 < d \leqslant 120$			

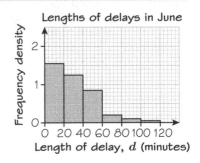

Lengths of delays in June

(2) The histogram shows information about the total rainfall, in millimetres, recorded at 60 weather stations during May.

Use the histogram to complete the frequency table and then estimate the mean rainfall.

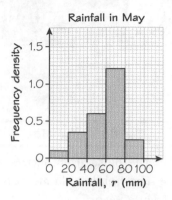

Rainfall in May

Rainfall, r (millimetres)	Frequency		
$0 < r \leqslant 20$			
$20 < r \leqslant 40$			
$40 < r \leqslant 60$			
$60 < r \leqslant 80$			
$80 < r \leqslant 100$			

(3) The histogram shows information about the waiting times during a clinic at a doctors' surgery.
Use the histogram to estimate the mean.

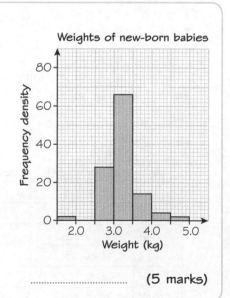

Waiting times

Hint Use the histogram to draw a frequency table.

Exam-style question

(4) The histogram shows the weights of babies born in a maternity ward during the first week of September.
Estimate the mean weight of the babies.

Give your answer to 3 significant figures.

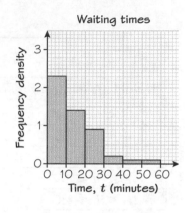

Weights of new-born babies

(5 marks)

Reflect How do you work out the number you divide the total by to estimate the mean?

Practise the methods

Answer this question to check where to start.

Check up

The table shows the masses of fish a fisherman caught.

Tick the correct histogram for the masses of the fish.

Mass, m (grams)	Frequency
$0 < m \leq 500$	12
$500 < m \leq 1000$	27
$1000 < m \leq 1500$	48
$1500 < m \leq 2000$	19

A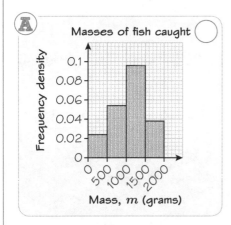

B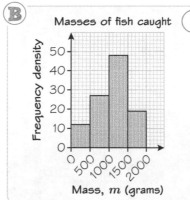

C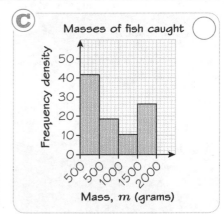

If you ticked A, go to Q2.

If you ticked B or C, go to Q1 for more practice.

(1) Work out the frequency densities for the information in each table.

a

Height, h (cm)	Frequency	
$0 < h \leq 20$	14	
$20 < h \leq 40$	23	
$40 < h \leq 60$	15	
$60 < h \leq 80$	8	

b

Mass, m (grams)	Frequency	
$0 < m \leq 100$	12	
$100 < m \leq 200$	47	
$200 < m \leq 300$	36	
$300 < m \leq 400$	14	

Exam-style question

(2) The histogram shows some information about the distances, in miles, travelled by people in an office to get to work.

8 people travel 5 miles or less to get to work.

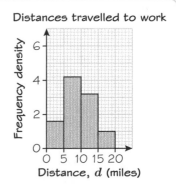

a Write down the modal class for the distances travelled to work. .... (1 mark)

b Work out how many people travelled over 10 miles to get to work.

.... (2 marks)

c Work out how many people were included in this survey.

.... (2 marks)

Problem-solve!

(1) Callum records the ages of the members of his running club in this table.

He wants to take a sample of 50 members.

Age, a	Frequency		
$15 < a \leq 30$	154		
$30 < a \leq 45$	171		
$45 < a \leq 60$	98		
$60 < a \leq 75$	3		

State whether Callum should take a random sample or a stratified sample. ....

Give a reason for your answer.

.... **(3 marks)**

(2) Draw a histogram using Callum's information from Q1.

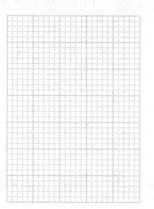

(3 marks)

(3) The table and the histogram show information about the masses of some apples.

Mass, m (grams)	Frequency		
$100 < w \leq 130$	6		
$130 < w \leq 160$	8		
$160 < w \leq 190$	20		
$190 < w \leq 220$			
$220 < w \leq 250$	5		

a Use the histogram to complete the table. **(2 marks)**

b Use the table to complete the histogram.

Masses of apples

Frequency density

0.8
0.6
0.4
0.2
0

100 130 160 190 220 250

Mass, m (grams) **(2 marks)**

c Estimate the mean mass of the apples.

Give your answer to 1 decimal place. .... **(2 marks)**

Now that you have completed this unit, how confident do you feel?

(1) Sampling

(2) Drawing histograms

(3) Estimating the mean from a histogram

② Indices and recurring decimals

This unit will help you to convert between recurring decimals and fractions and to calculate using fractional and negative indices.

Key points

A recurring decimal is a number where the digits after the decimal point keep repeating forever.

All recurring decimals can be written as fractions.

These **skills boosts** will help you to convert recurring decimals to fractions and use the laws of indices for fractional and negative indices.

> **①** Converting recurring decimals to fractions
>
> **②** Calculating using fractional and negative indices

You might have already done some work on fractions, decimals and indices. Before starting the first skills boost, rate your confidence with these questions.

① Write 0.454 545 454... as a fraction. Give your answer in its simplest form.

② Evaluate $27^{-\frac{2}{3}}$

How confident are you?

Converting recurring decimals to fractions

0.333 333... is a recurring decimal and can be written as $0.\dot{3}$.
0.345 345... is a recurring decimal and can be written as $0.\dot{3}4\dot{5}$.

Guided practice

Write $0.\dot{4}$ as a fraction.

_____

Form a simple equation by calling the recurring decimal x.

$x = 0.444\ 444...$

Multiply both sides by 10 to form another equation which moves the sequence one place to the left.

$10x = 4.444\ 444...$

Eliminate the recurring part of the decimal by subtracting the first equation, from the second one.

$10x - x = $ ....

Solve the equation and simplify if possible.

$x = $ ....

(1) Write each recurring decimal as a fraction in its simplest form.

    a $0.777\ 777...$ **b** $0.\dot{2}$ **c** $0.\dot{1}\dot{3}$

....

    d $0.5\dot{4}$ **e** $0.285\ 285\ 285...$ **f** $0.\dot{7}5\dot{6}$

....

Hint Multiply by 10 if 1 digit recurs, multiply by 100 if 2 digits recur, multiply by 1000 if 3 digits recur.

(2) Write each recurring decimal as a fraction in its simplest form.

    a $1.333\ 333...$ **b** $4.\dot{7}\dot{8}$ **c** $6.\dot{1}3\dot{5}$

....

    d $0.5\dot{1}$ **e** $0.1\dot{6}$ **f** $0.4\dot{8}\dot{6}$

....

Hint You must get two equations in x where the recurring part after the decimal point is exactly the same.

Exam-style question

(3) Express the recurring decimal $0.3\dot{8}$ as a fraction.
Give your answer in its simplest form.

.... **(3 marks)**

Reflect Are there any recurring decimals that you need to multiply by a number that is not a multiple of 10?

2 Calculating using fractional and negative indices

A fractional index tells you to find a root of the number, for example $4^{\frac{1}{2}} = \pm\sqrt{4} = \pm 2$.
A negative index tells you to work out the reciprocal, for example $5^{-2} = \frac{1}{5^2} = \frac{1}{25}$.

Worked exam question

Guided practice

Evaluate
a $64^{\frac{1}{3}}$ **b** $(10^2)^{\frac{1}{2}}$ **c** $32^{-\frac{2}{5}}$

---

a $64^{\frac{1}{3}} = \sqrt[3]{64} = $ ....

b $(10^2)^{\frac{1}{2}} = 10^{2\times\frac{1}{2}}$

$\qquad = 10$ ....

$\qquad = $ ....

For negative indices work out the reciprocal.

c $32^{-\frac{2}{5}} = \dfrac{1}{32^{\frac{2}{5}}}$

$\qquad = \dfrac{1}{\left(\sqrt[5]{32}\right)^2}$

$\qquad = \dfrac{1}{\underset{\cdots\cdots}{2}} = $ ....

A number to the power $\frac{1}{3}$ is the cube root of the number.

$(10^2)^{\frac{1}{2}} = 10^{2\times\frac{1}{2}}$
using the rule $(x^m)^n = x^{m\times n}$ or x^{mn}
$2 \times \frac{1}{2} = \frac{2}{1} \times \frac{1}{2} = \frac{2}{2} = 1$

$32^{\frac{2}{5}} = 32^{\frac{1}{5}\times 2} = \left(32^{\frac{1}{5}}\right)^2$ using the rule
$(x^m)^n = x^{m\times n}$ or x^{mn}

A number to the power $\frac{1}{5}$ is the fifth root of the number, so $32^{\frac{2}{5}} = \left(\sqrt[5]{32}\right)^2$

---

(1) Evaluate

  a $25^{\frac{1}{2}}$ **b** $81^{\frac{1}{2}}$ **c** $144^{\frac{1}{2}}$ **d** $8^{\frac{1}{3}}$

  e $1000^{\frac{1}{3}}$ **f** $225^{\frac{1}{2}}$ **g** $125^{\frac{1}{3}}$ **h** $169^{\frac{1}{2}}$

(2) Evaluate

  a $16^{\frac{1}{4}}$ **b** $100000^{\frac{1}{5}}$ **c** $81^{\frac{1}{4}}$ **d** $32^{\frac{1}{5}}$

(3) Evaluate

  a $\left(\dfrac{1}{100}\right)^{\frac{1}{2}}$ **b** $\left(\dfrac{27}{64}\right)^{\frac{1}{3}}$ **c** $\left(\dfrac{25}{64}\right)^{\frac{1}{2}}$ **d** $\left(\dfrac{27}{1000}\right)^{\frac{1}{3}}$

(4) Work out the missing indices.

  a $49^{\square} = 7$ **b** $64^{\square} = 4$ **c** $36^{\square} = 6$ **d** $128^{\square} = 2$

(5) Evaluate

  a $(4^2)^{\frac{1}{2}}$ **b** $(7^2)^{\frac{1}{2}}$ **c** $(3^{\frac{1}{2}})^2$ **d** $(10^2)^{\frac{1}{2}}$

6 Evaluate

a $(2^3)^{\frac{1}{3}}$ **b** $(5^3)^{\frac{1}{3}}$ **c** $(3^{\frac{1}{3}})^3$ **d** $(4^3)^{\frac{1}{3}}$

....

7 Work out the value of

a $(5^4)^{\frac{1}{4}}$ **b** $(10^{\frac{1}{5}})^5$ **c** $(7^6)^{\frac{1}{6}}$ **d** $(9^{\frac{1}{10}})^{10}$

....

8 Work out the missing indices.

a $(10^{\frac{1}{2}})^{\square} = 10$ **b** $(3^4)^{\square} = 3$ **c** $(7^{\square})^{\frac{1}{3}} = 7$ **d** $(8^{\square})^5 = 8$

....

9 Work out

a $27^{\frac{2}{3}}$ **b** $64^{\frac{3}{2}}$ **c** $64^{\frac{2}{3}}$ **d** $125^{\frac{2}{3}}$

....

e $81^{\frac{3}{4}}$ **f** $32^{\frac{3}{5}}$ **g** $36^{\frac{3}{2}}$ **h** $1000^{\frac{2}{3}}$

....

10 Calculate

a $16^{-\frac{1}{2}}$ **b** $125^{-\frac{1}{3}}$ **c** $81^{-\frac{1}{4}}$ **d** $121^{-\frac{1}{2}}$

....

11 Evaluate

a $8^{-\frac{2}{3}}$ **b** $64^{-\frac{2}{3}}$ **c** $9^{-\frac{3}{2}}$ **d** $81^{-\frac{3}{4}}$

....

e $1000^{-\frac{2}{3}}$ **f** $4^{-\frac{3}{2}}$ **g** $343^{-\frac{2}{3}}$ **h** $32^{-\frac{3}{5}}$

....

Exam-style question

12 Evaluate

a $16^{\frac{1}{2}}$ .... **(1 mark)**

b $(5^2)^{\frac{1}{2}}$ .... **(1 mark)**

c $27^{-\frac{2}{3}}$ .... **(1 mark)**

Reflect   Use your answers from Q6 to Q9 to help you simplify $(x^n)^{\frac{1}{n}}$.

Practise the methods

Answer this question to check where to start.

Check up

Tick the correct fraction for the recurring decimal 0.222...

 A $\frac{2}{10}$ ◯   **B** $\frac{222}{1000}$ ◯   **C** $\frac{0.2}{10}$ ◯   **D** $\frac{2}{9}$ ◯

If you ticked D, go to Q3.   If you ticked A, B or C go to Q1 for more practice.

1. Write each decimal as a fraction.
   a 0.3   b 0.7   c 0.33   d 0.373

2. Write each recurring decimal as a fraction in its simplest form.
   a 0.111 111...   b 0.$\dot{2}\dot{4}$ c 0.$\dot{4}0\dot{1}$

Exam-style question

3. Express the recurring decimal 0.7$\dot{2}$ as a fraction.
   Give your answer in its simplest form. .... (3 marks)

4. Work out
   a $\sqrt{36}$ b $\sqrt[3]{125}$ c $\sqrt[4]{81}$ d $\sqrt[4]{10\,000}$

5. Evaluate
   a $36^{\frac{1}{2}}$ b $125^{\frac{1}{3}}$ c $81^{\frac{1}{4}}$ d $10\,000^{\frac{1}{4}}$

6. What do you notice about your answers to Q4 and Q5?

7. Match the expressions with indices to the correct values.
   $16^{\frac{1}{2}}$ $16^{-\frac{1}{2}}$ $4^{\frac{3}{2}}$ $4^{-\frac{3}{2}}$ $64^{\frac{2}{3}}$ $64^{-\frac{2}{3}}$

   16   8 $\frac{1}{16}$ 4 $\frac{1}{8}$ $\frac{1}{4}$

Exam-style question

8. a Evaluate $100^{\frac{1}{2}}$ .... (1 mark)
   b Work out the value of $\left(\frac{64}{125}\right)^{\frac{2}{3}}$ .... (2 marks)
   c Calculate the value of x when $(5^x)^{\frac{1}{4}} = 5$ .... (2 marks)

Problem-solve!

① Express the recurring decimal $0.21\dot{3}$ as a fraction.
Give your answer in its simplest form.

.... (3 marks)

② **a** Write down the value of $4^{\frac{1}{2}}$

.... (1 mark)

b Find the value of $8^{\frac{2}{3}}$

.... (1 mark)

c Work out $\left(\dfrac{16}{25}\right)^{-\frac{3}{2}}$

.... (2 marks)

③ **a** Write down the value of $27^{\frac{1}{3}}$

.... (1 mark)

b Find the value of $49^{-\frac{1}{2}}$

.... (1 mark)

c Evaluate $32^{-\frac{2}{5}}$

.... (2 marks)

④ **a** Write down the value of $(x^3)^{\frac{1}{3}}$

.... (1 mark)

b Work out the value of $100^{\frac{1}{2}} \times 125^{-\frac{1}{3}}$

.... (2 marks)

⑤ Write these numbers in order of size.
Start with the smallest number.

$25^{\frac{1}{2}}$ 　　　　　　 0.5 　　　　　　 -5 　　　　　　 $25^{-\frac{1}{2}}$

.... (2 marks)

Now that you have completed this unit, how confident do you feel?

❶ Converting recurring decimals to fractions

❷ Calculating using fractional and negative indices

③ Surds

This unit will help you to simplify expressions including surds and to rationalise denominators.

AO1 Fluency check

① Expand the brackets and simplify.

a $(x + 2)(x - 3)$ .... **b** $(x - 5)(x + 5)$ ....

c $(x - 4)(2x - 7)$ .... **d** $(2x + 5)(x - 3)$ ....

② Work out

a $\dfrac{2}{3} \times \dfrac{3}{4}$ .... **b** $\dfrac{4}{5} \times \dfrac{5}{8}$ .... **c** $\dfrac{7}{9} \times \dfrac{3}{3}$ .... **d** $\dfrac{3}{8} \times \dfrac{5}{5}$ ....

③ Work out

a $\left(\sqrt{25}\right)^2$ .... **b** $\left(\sqrt{4}\right)^2$ .... **c** $\left(\sqrt{7}\right)^2$ .... **d** $\left(\sqrt{3}\right)^2$ ....

④ Number sense

Write each number as a product of two numbers, where one of the numbers is a square number.

a 45 .... **b** 75 .... **c** 80 .... **d** 147 ....

Key points

A surd is the square root of a number that cannot be simplified, for example, $\sqrt{2}$ and $\sqrt{5}$ are surds. $\sqrt{4} = 2$ so $\sqrt{4}$ is not a surd.

The square root of a number, squared, is equal to the number, for example $\left(\sqrt{x}\right)^2 = x$

These **skills boosts** will help you to simplify surds and to rationalise denominators.

① Simplifying surds ② Expanding brackets involving surds ③ Rationalising denominators ④ Simplifying algebraic expressions involving surds

You might have already done some work on surds. Before starting the first skills boost, rate your confidence with these questions.

① Simplify $\sqrt{12}$.

② Write $(2 - \sqrt{3})^2$ in the form $a + b\sqrt{3}$.

③ Rationalise the denominator of $\dfrac{2}{\sqrt{5}}$

④ Expand and simplify $(1 - \sqrt{x})^2$.

How confident are you?
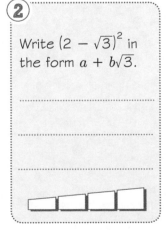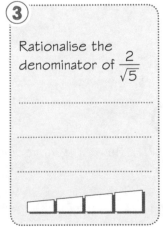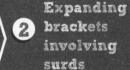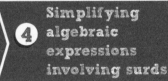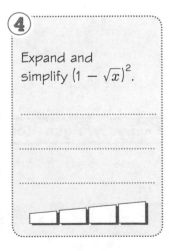

Simplifying surds

Useful rules for surds are $\sqrt{ab} = \sqrt{a}\sqrt{b}$ and $\sqrt{\dfrac{a}{b}} = \dfrac{\sqrt{a}}{\sqrt{b}}$

For example

$\sqrt{4 \times 9} = \sqrt{36} = 6$ and $\sqrt{4} \times \sqrt{9} = 2 \times 3 = 6$ $\sqrt{\dfrac{100}{4}} = \sqrt{25} = 5$ and $\dfrac{\sqrt{100}}{\sqrt{4}} = \dfrac{10}{2} = 5$

Guided practice

Simplify $\sqrt{12}$.

Write 12 as the product of two numbers where one of the numbers is a square number.
(So do not use 6×2 or 12×1.)

$\sqrt{12} = \sqrt{4 \times 3}$

Use the rule $\sqrt{ab} = \sqrt{a}\sqrt{b}$

$\quad = \sqrt{4} \times \sqrt{\rule{1cm}{0.4pt}}$

Work out $\sqrt{4}$ and simplify.

$\quad = \rule{1cm}{0.4pt} \times \sqrt{\rule{1cm}{0.4pt}}$

$\quad = \rule{1cm}{0.4pt}\sqrt{\rule{1cm}{0.4pt}}$

① Simplify

　a $\sqrt{45}$　　　　　**b** $\sqrt{75}$　　　　　**c** $\sqrt{48}$　　　　　**d** $\sqrt{80}$

....　........................　........................　........................

　e $\sqrt{200}$　　　　　**f** $\sqrt{147}$　　　　　**g** $\sqrt{50}$　　　　　**h** $\sqrt{32}$

....　........................　........................　........................

② Simplify

　a $2\sqrt{12}$　　　　　**b** $3\sqrt{28}$　　　　　**c** $5\sqrt{27}$　　　　　**d** $2\sqrt{18}$

....　........................　........................　........................

③ Simplify

　a $\sqrt{\dfrac{7}{9}}$　　　　　**b** $\sqrt{\dfrac{11}{25}}$　　　　　**c** $\sqrt{\dfrac{12}{25}}$　　　　　**d** $\sqrt{\dfrac{28}{49}}$

....　........................　........................　........................

Hint Use the rule $\sqrt{\dfrac{a}{b}} = \dfrac{\sqrt{a}}{\sqrt{b}}$

Exam-style question

④ Write $\sqrt{20}$ in the form $k\sqrt{5}$, where k is an integer. .... **(2 marks)**

Reflect Write a set of instructions for yourself on how to simplify surds.

2 Expanding brackets involving surds

Guided practice

Expand $(2 - \sqrt{3})(1 + \sqrt{3})$.

Give your answer in the form $a + b\sqrt{3}$, where a and b are integers.

Multiply every term in the first bracket by every term in the second bracket.
Expand the brackets.

$(2 - \sqrt{3})(1 + \sqrt{3}) = 2(1 + \sqrt{3}) - \sqrt{3}(1 + \sqrt{3})$

Collect like terms.

$$= 2 + \underbrace{\rule{1.5cm}{0.4pt} - \rule{1.5cm}{0.4pt}} - 3$$

$$= -1 + \rule{1.5cm}{0.4pt}$$

(1) Expand and simplify

a $(1 + \sqrt{3})(1 - \sqrt{3})$ **b** $(3 - \sqrt{5})(2 + \sqrt{5})$ **c** $(1 + \sqrt{2})(3 + \sqrt{2})$

d $(2 + \sqrt{11})(1 + \sqrt{11})$ **e** $(4 - \sqrt{7})(1 - \sqrt{7})$ **f** $(3 + \sqrt{5})(1 - \sqrt{5})$

(2) Expand and simplify

a $(1 + \sqrt{2})^2$ **b** $(2 - \sqrt{5})^2$ **c** $(3 + \sqrt{7})^2$ **d** $(1 - \sqrt{3})^2$

Hint $(1 + \sqrt{2})^2 = (1 + \sqrt{2})(1 + \sqrt{2})$

Exam-style questions

(3) Write $(3 - \sqrt{3})^2$ in the form $a + b\sqrt{3}$, where a and b are integers.

.... (2 marks)

(4) Work out the value of $(\sqrt{12} + \sqrt{3})^2$.

.... (2 marks)

(5) Expand $(3 + \sqrt{2})(1 - \sqrt{2})$.

Give your answer in the form $a + b\sqrt{2}$, where a and b are integers.

.... (2 marks)

Reflect What alternative methods do you know for expanding a pair of brackets?

3 Rationalising denominators

Rationalising the denominator means removing the surd from the bottom of a fraction.

Guided practice

Rationalise the denominator of $\dfrac{4}{\sqrt{5}}$.

$$\frac{4}{\sqrt{5}} = \frac{4}{\sqrt{5}} \times \frac{\sqrt{5}}{\sqrt{5}}$$

$$= \frac{4 \times \text{.........}}{\text{..............}} = \text{..............}$$

$\dfrac{\sqrt{5}}{\sqrt{5}} = 1$

Multiplying by 1 does not change the value.

① Rationalise the denominators.

a $\dfrac{1}{\sqrt{2}}$ .... **b** $\dfrac{1}{\sqrt{7}}$ .... **c** $\dfrac{1}{\sqrt{3}}$ .... **d** $\dfrac{1}{\sqrt{5}}$ ....

e $\dfrac{1}{\sqrt{11}}$ .... **f** $\dfrac{1}{\sqrt{13}}$ .... **g** $\dfrac{1}{\sqrt{17}}$ .... **h** $\dfrac{1}{\sqrt{19}}$ ....

② Rationalise the denominators.

a $\dfrac{3}{\sqrt{2}}$ .... **b** $\dfrac{2}{\sqrt{3}}$ .... **c** $\dfrac{3}{\sqrt{5}}$ .... **d** $\dfrac{4}{\sqrt{3}}$ ....

e $\dfrac{5}{\sqrt{7}}$ .... **f** $\dfrac{2}{\sqrt{5}}$ .... **g** $\dfrac{4}{\sqrt{7}}$ .... **h** $\dfrac{6}{\sqrt{11}}$ ....

③ Rationalise the denominators and simplify if possible.

a $\dfrac{4}{\sqrt{2}}$ .... **b** $\dfrac{3}{\sqrt{3}}$ .... **c** $\dfrac{2}{\sqrt{2}}$ .... **d** $\dfrac{6}{\sqrt{3}}$ ....

e $\dfrac{14}{\sqrt{7}}$ .... **f** $\dfrac{5}{\sqrt{5}}$ .... **g** $\dfrac{7}{\sqrt{7}}$ .... **h** $\dfrac{22}{\sqrt{11}}$ ....

Exam-style question

④ Rationalise the denominator of $\dfrac{10}{\sqrt{5}}$.

Give your answer in its simplest form. .... **(2 marks)**

Reflect

What do you multiply $\dfrac{a}{\sqrt{b}}$ by to rationalise the denominator?

Simplifying algebraic expressions involving surds

All of the methods covered in this unit can be used to simplify algebraic expressions.

Guided practice .... | **Worked exam question**

a Expand and simplify $(x + \sqrt{y})^2$.

b Rationalise the denominator of $\dfrac{1}{\sqrt{x}}$.

_____

Multiply every term in the first bracket by every term in the second bracket.

Expand the brackets.

a $\left(x + \sqrt{y}\right)^2 = x\left(x + \sqrt{y}\right) + \sqrt{y}\left(x + \sqrt{y}\right)$

Collect like terms.

$$= x^2 + \text{.............} + \text{.............} + y$$

$$= \text{..}$$

b $\dfrac{1}{\sqrt{x}} = \dfrac{1}{\sqrt{x}} \times \dfrac{\sqrt{x}}{\sqrt{x}} = $ ....

$\dfrac{\sqrt{x}}{\sqrt{x}} = 1$

Multiplying by 1 does not change the value.

① Expand and simplify

   a $(a + \sqrt{b})(a - \sqrt{b})$ **b** $(m - \sqrt{n})^2$ **c** $(p + \sqrt{q})(p - \sqrt{q})$

   ....

② Simplify fully

   a $(a - 2\sqrt{b})(2a + 3\sqrt{b})$ **b** $(j + 4\sqrt{k})^2$ **c** $(2c + \sqrt{d})(c - 2\sqrt{d})$

   ....

③ Rationalise the denominators.

   a $\dfrac{1}{\sqrt{a}}$ ....   **b** $\dfrac{2}{\sqrt{p}}$ ....   **c** $\dfrac{5}{\sqrt{t}}$ ....

④ Rationalise the denominators.

   a $\dfrac{1}{\sqrt{2b}}$ ....   **b** $\dfrac{6}{\sqrt{3u}}$ ....   **c** $\dfrac{5}{\sqrt{5e}}$ ....

Exam-style question

⑤ Simplify fully $(\sqrt{a} + \sqrt{9b})(\sqrt{a} - 3\sqrt{b})$.

.... (3 marks)

Reflect Is it more difficult to apply the rules for surds to numerical expressions or to algebraic expressions? Explain your answer.

Practise the methods

Answer this question to check where to start.

Check up

Tick the best product for simplifying $\sqrt{300}$.

A $\sqrt{150 \times 2}$ ◯ **B** $\sqrt{300 \times 1}$ ◯ **C** $\sqrt{100 \times 3}$ ◯ **D** $\sqrt{25 \times 12}$ ◯

If you ticked C, finish simplifying $\sqrt{300}$ then go to Q2.

If you ticked A, B or D go to Q1 for more practice.

① **a** Tick the best product for simplifying $\sqrt{128}$.

◯ $\sqrt{128 \times 1}$ ◯ $\sqrt{64 \times 2}$ ◯ $\sqrt{32 \times 4}$ ◯ $\sqrt{16 \times 8}$

b Simplify $\sqrt{128}$. ....

② Simplify

a $\sqrt{125}$ **b** $\sqrt{162}$ **c** $3\sqrt{32}$ **d** $5\sqrt{18}$

③ Rationalise the denominators.
Give your answers in their simplest form.

a $\dfrac{3}{\sqrt{5}}$ **b** $\dfrac{12}{\sqrt{3}}$ **c** $\dfrac{10}{\sqrt{2}}$ **d** $\dfrac{k}{\sqrt{k}}$

④ Expand and simplify

a $(1 + \sqrt{5})(1 - \sqrt{5})$ **b** $(2 + \sqrt{11})(3 + \sqrt{11})$ **c** $(2a + \sqrt{b})^2$

Exam-style questions

⑤ Write $5\sqrt{12}$ in the form $k\sqrt{3}$, where k is an integer.

.... (2 marks)

⑥ Write $(5 + \sqrt{5})^2$ in the form $a + b\sqrt{5}$, where a and b are integers.

.... (2 marks)

Problem-solve!

(1) Write $\sqrt{63}$ in the form $k\sqrt{7}$, where k is an integer.

.... (2 marks)

(2) Write $(3 - \sqrt{3})^2$ in the form $a + b\sqrt{3}$, where a and b are integers.

.... (2 marks)

(3) Rationalise the denominator of $\dfrac{9}{\sqrt{3}}$.

Give your answer in its simplest form.

.... (2 marks)

(4) Show that $\dfrac{1 + \sqrt{3}}{\sqrt{2}}$ can be written as $\dfrac{\sqrt{2} + \sqrt{6}}{2}$.

.... (2 marks)

(5) Expand and simplify $(3 + \sqrt{2})^2 - (3 - \sqrt{2})^2$.

.... (2 marks)

(6) Simplify fully $(\sqrt{x} + \sqrt{4y})(\sqrt{x} - 2\sqrt{y})$.

.... (3 marks)

(7) The perimeter of a square is $\sqrt{60}$ cm.

Work out the area of the square.

Give your answer in its simplest form.

.... (3 marks)

Now that you have completed this unit, how confident do you feel?

1 Simplifying surds

2 Expanding brackets involving surds

3 Rationalising denominators

4 Simplifying algebraic expressions involving surds

④ Cumulative frequency

This unit will help you to draw and interpret cumulative frequency graphs and box plots and to compare distributions.

AO1 Fluency check

① Work out the median for the set of data.

8   10   9   8   7   12   10

② Work out the median for the set of data.

34   38   33   35   33   36

③ Work out the range for the set of data in Q2.

④ **Number sense**

Work out

a $64 + 43 + 28 + 26 + 52$ ....   **b** $121 + 65 + 39 + 79$ ....

Key points

Cumulative means increasing in quantity as more is added.	Frequency means how many times something occurs.

These **skills boosts** will help you to draw cumulative frequency graphs and box plots, interpret cumulative frequency graphs and box plots and compare distributions.

> ① Constructing cumulative frequency graphs   ② Drawing and interpreting box plots   ③ Comparing distributions

You might have already done some work on cumulative frequency and box plots. Before starting the first skills boost, rate your confidence with these questions.

① Work out the cumulative frequencies for the information in the table.

Height, h (cm)	Frequency
$0 < h \leq 10$	5
$10 < h \leq 20$	15
$20 < h \leq 30$	23
$30 < h \leq 40$	12
$40 < h \leq 50$	3

② To draw a box plot you need to know five pieces of information.

You need the lowest and highest values in the data set, and the lower and upper quartiles. What other piece of information do you need?

③ In a science test, the interquartile range for the boys' results is 42 marks. For the girls' results it is 35 marks.

Compare the boys' and girls' results using the interquartile range.

How confident are you?

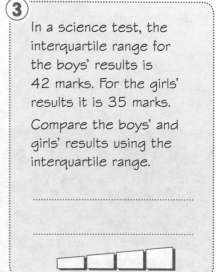

1 Constructing cumulative frequency graphs

On a cumulative frequency graph, cumulative frequency is on the vertical axis.

The frequency table shows the heights of 80 bean plants.

Draw a cumulative frequency graph for the heights of the plants.

Height, h (cm)	Frequency
$0 < h \leqslant 10$	3
$10 < h \leqslant 20$	21
$20 < h \leqslant 30$	35
$30 < h \leqslant 40$	16
$40 < h \leqslant 50$	5

Height, h (cm)	Cumulative frequency	Explanation
$0 < h \leqslant 10$	3	3 plants are up to 10 cm tall
$0 < h \leqslant 20$	3 + 21 = 24	24 plants are up to 20 cm tall
$0 < h \leqslant 30$	3 + 21 + 35 = ...	.... plants are up to 30 cm tall
$0 < h \leqslant 40$		
$0 < h \leqslant 50$		

Draw a cumulative frequency table to show your working for the cumulative frequencies.

Use the information in the Explanation column to plot the cumulative frequency graph.

Each cumulative frequency is the total of the frequency for the class and all the previous frequencies.

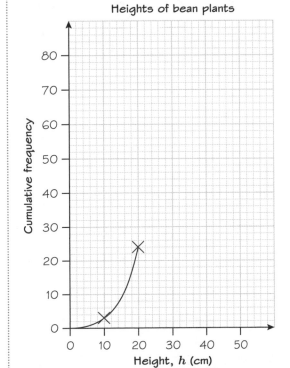

Heights of bean plants

Join the points using a smooth curve.

① The frequency table shows the masses of parcels which a post office delivers in one week.

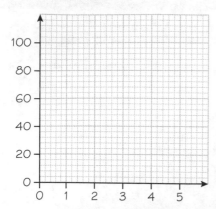

Mass, m (kg)	Frequency		
$0 < m \leqslant 1$	21		
$1 < m \leqslant 2$	32		
$2 < m \leqslant 3$	22		
$3 < m \leqslant 4$	17		
$4 < m \leqslant 5$	8		

Draw a cumulative frequency graph for the masses of the parcels.

Exam-style question

② The table shows some information about the times, in minutes, 60 people took to get to work.

Time, t (minutes)	Frequency		
$0 < t \leqslant 10$	3		
$10 < t \leqslant 20$	12		
$20 < t \leqslant 30$	19		
$30 < t \leqslant 45$	21		
$45 < t \leqslant 60$	5		

Draw a cumulative frequency graph for the time taken to get to work.

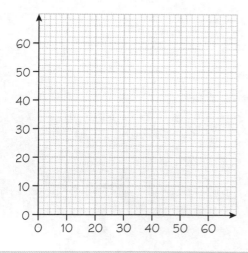

(3 marks)

Reflect   Explain where you plot the points on a cumulative frequency graph.

2 Drawing and interpreting box plots

A box plot shows the median, the quartiles and the maximum and minimum values.

Guided practice

Here are the ages, in years, of 15 women in a running club.

16, 17, 17, 19, 20, 24, 25, 25, 26, 28, 30, 31, 31, 35, 57

a Draw a box plot for this information.

b Identify any outliers in the data.

---

a Median = 25

Lower quartile = 19

Upper quartile = ...

The median splits the data set in half when the numbers are in numerical order.

The quartiles split the data set into quarters.

Ages of women in a running club

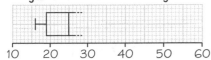

```
10   20   30   40   50   60
```

The lower and upper quartiles make the box of the box plot. The median is inside the box.

b Interquartile range (IQR) = ...

1.5 × IQR = ...

Lower quartile − 1.5 × IQR = ...

Upper quartile + 1.5 × IQR = ...

Outliers: ....

IQR
= upper quartile − lower quartile

An outlier can be defined as a data value more than 1.5 × IQR below the lower quartile or more than 1.5 × IQR above the upper quartile.

---

① Hwan played 15 games of basketball.
Here are the points he scored in each game.

6, 17, 18, 19, 19, 20, 21, 22, 22, 23, 23, 24, 26, 27, 27

a Draw a box plot for this information.

b Identify any outliers in the data. ....

② Here are the weekly wages of 11 members of staff in a company.

148, 225, 275, 320, 340, 350, 350, 415, 450, 450, 650

a Draw a box plot for this information.

b Identify any outliers in the data. ....

③ The table gives information about the times, in minutes, 150 students revised for an exam.

Least	Lower quartile	Median	Upper quartile	Most
15	20	45	60	90

Draw a box plot for this information.

④ The table shows information about the heights, in metres, of 50 sunflower plants.

Shortest	Lower quartile	Median	Upper quartile	Tallest
1.65	1.9	2.1	2.35	2.5

Draw a box plot for this information.

⑤ The box plot shows information about the midday temperatures on every Monday last year.

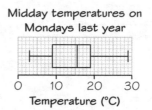

Midday temperatures on Mondays last year

Temperature (°C)

a Write down the median midday temperature on Mondays last year. ....

b Write down the maximum and minimum midday temperatures on Mondays last year. ....

c Work out the interquartile range.

....

Exam-style question

⑥ The box plot shows information about the masses of 100 children.

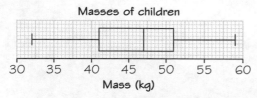

Masses of children

Mass (kg)

a Write down the median. .... (1 mark)

b Work out the interquartile range. .... (2 marks)

There are 100 children in the group.

c Work out the number of children in the group who have a mass of 51 kg or more.

.... (2 marks)

Reflect Look at your box plots. What information do you get from a box plot?

3 Comparing distributions

You can use the median and interquartile range (IQR) from cumulative frequency graphs and box plots to compare distributions.

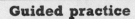

Guided practice

The box plot shows the temperatures in London in October.

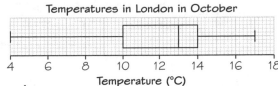

Temperatures in London in October

Temperature (°C)

a Write down the median temperature.

b Work out the IQR.

This box plot shows the temperatures in Glasgow in October.

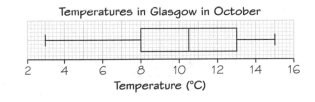

Temperatures in Glasgow in October

Temperature (°C)

c Compare the distribution of the temperatures in London with the distribution of the temperatures in Glasgow.

a Median = ... °C

b IQR = upper quartile − lower quartile = ... − .... = ... °C

c Median for Glasgow = ... °C

Work out the median and IQR to compare the distributions.

IQR for Glasgow = upper quartile − lower quartile = ... − .... = ... °C

The median temperature for London is ....

and for Glasgow is ...., so on average

the temperature for .... is higher.

The IQR for Glasgow is slightly greater than for

London, suggesting that the temperatures in

.... are slightly more spread out than the temperatures in ....

Median temperature for London = ☐
Median temperature for Glasgow = ☐
Compare to say which city is warmer.

Use the IQR to compare the spread of the distributions.

① The box plot shows the amounts of vitamin C in different fruits.

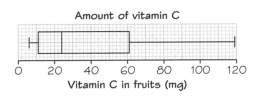

Amount of vitamin C

Vitamin C in fruits (mg)

a Write down the median amount of vitamin C. ....

b Work out the IQR. ....

This box plot shows the amounts of vitamin C in different vegetables.

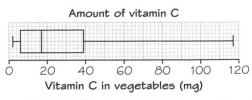

Amount of vitamin C

Vitamin C in vegetables (mg)

c Compare distribution of amounts of vitamin C in fruits with the distribution of amounts of vitamin C in vegetables.

....

Hint Refer to the context in your comparison. In this case, the context is the amounts of vitamin C.

(2) The back to back stem and leaf diagram shows the lengths of calls, to the nearest minute, that two call-centre employees took in a day.

Compare the distribution of the lengths of calls taken by the two employees.

Hint Use the median or mean and the IQR or range to compare the length of calls.

Lengths of calls

Employee A		Employee B
9 9 5 5 5 3 2 2	0	4 7
9 8 7 6 6 5 3 3 1 1	1	5
7 4	2	2 3 8 8
2	3	1 6 7
	4	2

Key
2|0 represents 2 minutes for employee A.
0|4 represents 4 minutes for employee B.

....

....

....

Exam-style question

(3) The cumulative frequency graph shows the patient waiting times, to the nearest minute, for a morning clinic in a GP surgery.

a Work out the median waiting time.

.... **(1 mark)**

b Work out the interquartile range.

.... **(2 marks)**

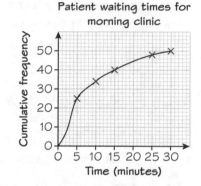

Patient waiting times for morning clinic

The box plot shows the patient waiting times for an afternoon clinic at the surgery.

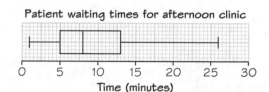

Patient waiting times for afternoon clinic

c Compare the two distributions. ....

....

....

.... **(2 marks)**

Reflect What information can you use to compare distributions?

Practise the methods

Answer this question to check where to start.

Check up

Tick the correct cumulative frequency graph for the information in the frequency table.

Height, h (cm)	Frequency
$70 \leqslant h < 80$	8
$80 \leqslant h < 90$	15
$90 \leqslant h < 100$	11
$100 \leqslant h < 110$	6

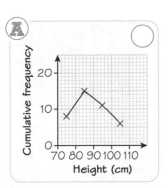

A ○ **B** ○ **C** ○ **D** ○

If you ticked D work out the median and the interquartile range. Then go to Q2.

If you ticked A, B or C go to Q1 for more practice.

---

① Complete the tables to show the coordinates you would use to plot a cumulative frequency graph for each frequency table.

a

Mass, m (kg)	Frequency	Coordinates
$40 < w \leqslant 50$	3	(50, 3)
$50 < w \leqslant 60$	7	(60, 10)
$60 < w \leqslant 70$	16	
$70 < w \leqslant 80$	10	
$80 < w \leqslant 90$	4	

b

Time, t (hours)	Frequency	Coordinates
$0 < t \leqslant 0.5$	3	(0.5, 3)
$0.5 < t \leqslant 1$	6	
$1 < t \leqslant 1.5$	14	
$1.5 < t \leqslant 2$	5	
$2 < t \leqslant 2.5$	2	

② Draw a cumulative frequency graph for each frequency table in Q1.

a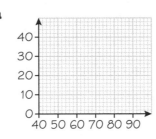

b

Exam-style question

③ The cumulative frequency graph gives information about the test results of two groups of students, group A and group B. Compare the test results of the students in group A with the test results of the students in group B.

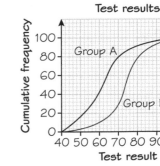

....

....

....

....

(4 marks)

Problem-solve!

1. The table shows information about the times taken by 100 people in a fun run.

Time, t (minutes)	Frequency
$10 < t \leq 20$	3
$20 < t \leq 30$	15
$30 < t \leq 40$	35
$40 < t \leq 50$	26
$50 < t \leq 60$	15
$60 < t \leq 70$	6

a Complete the cumulative frequency table for this information.

Time, t (minutes)	Cumulative frequency
$10 < t \leq 20$	
$10 < t \leq 30$	
$10 < t \leq 40$	
$10 < t \leq 50$	
$10 < t \leq 60$	
$10 < t \leq 70$	

(1 mark)

b Draw a cumulative frequency graph for your table.

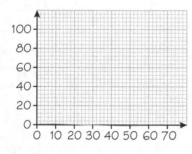

(2 marks)

c Use your graph to find an estimate for the median time. .... (1 mark)

d Use your graph to find an estimate for the number of people
who took longer than 53 minutes. .... (2 marks)

2. The table gives information about the masses of 80 parcels.

lowest	1.2 kg
highest	6.1 kg
lower quartile	2.6 kg
interquartile range	1.5 kg
median	3.6 kg

Draw a box plot for this information.

(3 marks)

③ Felix recorded the heart rate, in beats per minute, of each of 15 people.

He then asked the 15 people to walk up some stairs.

He recorded their heart rates again.

He showed his results in a back-to-back stem and leaf diagram.

Heart rates before and after walking up the stairs

Before		After	Key

Before		After
9	5	
8 8 7 5 3 2 0	6	5 7 9 9
8 7 5 4 1	7	1 5 6 8
5 2	8	0 4 7
	9	2 4 8
	10	3

Key

9 | 5 means 59 beats per minute for Before.

6 | 5 means 65 beats per minute for After.

Compare the heart rates of the people before they walked up the stairs with their heart rates after they walked up the stairs.

....

....

....

....

....

.... **(6 marks)**

④ The cumulative frequency graph gives information about the heights of two groups of children, group A and group B.

Compare the heights of the children in group A with the heights of the children in group B.

....

....

....

....

....

....

....

.... **(4 marks)**

Heights of children

Cumulative frequency

Group A

Group B

Height (cm)

Now that you have completed this unit, how confident do you feel?

① Constructing cumulative frequency graphs

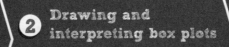
② Drawing and interpreting box plots

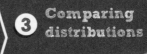
③ Comparing distributions

⑤ Probability

This unit will help you to work out how many outcomes there are for combined events and to understand and calculate the probabilities of dependent events and conditional probabilities.

AO1 Fluency check

① The probability of it raining tomorrow is 0.4.

Work out the probability of it not raining tomorrow.

_____

② The probability of rolling a 3 on a dice is $\frac{1}{6}$

The probability of spinning a 3 on a four-sided spinner is $\frac{1}{4}$

Sarah rolls the dice and spins the spinner.

Work out the probability of the dice and the spinner both landing on 3.

_____

③ **Number sense**

Work out

a 0.2×0.7 _____   **b** 0.3×0.1 _____   **c** $\frac{1}{3} \times \frac{2}{5}$ _____   **d** $\frac{3}{4} \times \frac{5}{6}$ _____

Key points

If one event depends on the outcome of another, the two events are dependent, for example picking two chocolates from a box without replacement.

Conditional probability is the probability of one event happening given that another event has happened.

These **skills boosts** will help you to understand and calculate the probabilities of dependent events and conditional probabilities and work out how many outcomes there are for combined events.

> ① **Combining events** > ② **Tree diagrams** > ③ **Venn diagrams for conditional probability**

You might have already done some work on the probabilities of combined events and dependent events and on conditional probabilities. Before starting the first skills boost, rate your confidence using these questions.

①

Benas buys an ice cream.

He has a choice of five flavours and three toppings.

Work out the number of possible combinations of ice cream and toppings.

②

There are eight cartons of juice in the fridge.

Five of the cartons contain apple juice and three contain orange juice.

Work out the probability that two cartons of juice taken at random are both apple juice.

③

A one-week survey of 120 students shows if they have a school dinner (D) or a packed lunch (P). 70 students have school dinners and packed lunches during the week. 80 students have school dinners. (a) Draw a Venn diagram. (b) Use your Venn diagram to work out $P(D \cap P \mid P)$.

How confident are you?

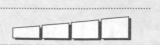

1 Combining events

If there are x ways one event can happen and y ways another event can happen, then there are $x \times y$ ways the two events can happen together.

Guided practice

A set menu offers a choice of four main courses and three desserts.

a Work out the number of combinations of main course and dessert.

b Work out the number of combinations if the menu has six main courses and four desserts.

a

A diagram can help you visualise the problem. Call the main courses M1, M2, M3 and M4. Call the desserts D1, D2 and D3.

Number of combinations = 4 × .... = ...

b Number of combinations = 6 × .... = ...

This method is called the product rule.

① A furniture shop sells four types of chairs, covered in five different fabrics.

   a Work out the number of combinations of chairs and fabrics. ....

   b Work out the number of combinations for three chairs and seven fabrics. ....

   c Work out the number of combinations for five chairs and three fabrics. ....

② A pet shop sells three different sizes of dog collar in six different colours.
   How many possible combinations of size and colour are there? ....

③ Two six-sided dice are rolled at the same time.
   How many outcomes are there altogether? ....

④ A five-sided spinner is spun twice.
   How many outcomes are there altogether? ....

⑤ Adele, Bala, Celia, Dan and Eden form a badminton team and hold a tournament against a team of seven of their parents.
   Each member of the team plays every parent, including their own.
   How many matches are there altogether? ....

Exam-style question

⑥ A school chooses a boy and a girl from Year 11 to represent it in a tennis tournament.
   There are nine boys and seven girls who would like to be chosen.
   Work out the number of combinations for the two representatives
   being a boy and a girl. .... **(2 marks)**

⑦ An m-sided dice is rolled and an n-sided spinner is spun.
   Write an expression for the number of outcomes for the dice and the spinner. ....

Reflect Explain how you work out the number of combinations for two events.

2 Tree diagrams

A tree diagram can help you work out the combined probabilities of more than one event. The probability that one event happens given that another event has already happened is a conditional probability.

Guided practice

Amir plays two tennis matches.

His probability of winning the first match is 0.65.

His probability of winning the second match having won the first is 0.75.

His probability of losing the second match having lost the first is 0.5.

Work out the probability that Amir wins exactly one match.

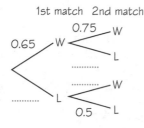

1st match   2nd match

P(exactly one win) = P(W, L) + P(L, W)

= .... × .... + ... × ....

= ....

Draw a probability tree diagram. Clearly label the probabilities given in the question.
The outcome of the second match depends on the outcome of the first match.
These events are dependent events.

If Amir wins the first match, the probability that he wins the second match is 0.75.
This is a conditional probability.

① There are eight counters in a bag.
Four are red and four are green.
Two counters are picked at random.
Work out the probability that the two counters are different colours.

Hint Draw a probability tree diagram.

....

② Zainab either cycles to work or takes the bus.
The probability that she cycles to work is 0.6.
If she cycles to work, the probability that she will be late is 0.05.
If she takes the bus to work, the probability that she will be late is 0.15.

a Complete the probability tree diagram.

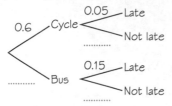

b Work out the probability that Zainab will not be late for work.   ....

③ In a school at lunchtime you can either have
a school dinner or take a packed lunch.
45% of the students are girls.
60% of the girls have school dinners.
85% of the boys take a packed lunch.
One student is selected at random.
What is the probability that this student
has a packed lunch?

....

④ For each pair of events, state if the events are independent or dependent.

   a Spinning a five-sided spinner, numbered 1 to 5, and rolling a dice.   ....

   b Dealing two cards from a pack of playing cards.   ....

   c Picking two sweets at random from a bag.   ....

   d Flipping a coin twice.   ....

⑤ There are 12 counters in a bag.
Three of the counters are red.
Four of the counters are blue. Five of the counters are green.
Mark takes a counter from the bag at random. He keeps the counter.
He then takes another counter from the bag at random.

   a Complete the probability tree diagram.

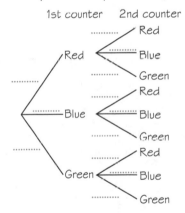

   b Work out the probability that Mark takes two different coloured counters
      from the bag.   ....

Exam-style question

⑥ There are nine girls and six boys in a drama group.
Ali is going to pick at random two members from
the drama group.
Work out the probability that Ali will pick
two boys or two girls.

....   **(4 marks)**

Reflect   Explain how you know if events are independent or dependent. You could use your
      answer to Q4 to help you.

3 Venn diagrams for conditional probability

P ∩ Q means the intersection of sets P and Q.

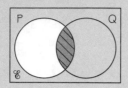

P ∪ Q means the union of sets P and Q.

P′ means the elements not in set P.

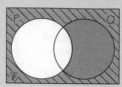

Guided practice

Worked exam question

An ice cream parlour sells a three-scoop sundae.

Customers can choose between vanilla, chocolate and strawberry ice cream.

One week the ice cream parlour sold 250 sundaes.

126 of the sundaes were one scoop of each flavour. 139 were vanilla and strawberry.

145 were vanilla and chocolate. 131 were strawberry and chocolate.

177 were chocolate. 173 were strawberry.

a Draw a Venn diagram to show this information.

A customer is selected at random.

b Given that the customer chooses vanilla ice cream, find the probability that this customer also chooses chocolate.

---

126 sundaes had all three flavours so write 126 in the overlap of all three circles. 139 sundaes were vanilla and strawberry. 126 of these are already shown, so you only need to fill in 139 − 126 = ☐. Work out the numbers for vanilla and chocolate, and for chocolate and strawberry, in the same way.

a

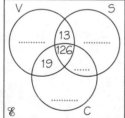

177 sundaes are chocolate, so all of the numbers in the circle for chocolate must total 177.

V ∩ C | V means the elements in the intersection of the sets V and C, given that V happens.

b $P(V \cap C \mid V) = \dfrac{145}{\underline{}}$

V ∩ C is the total of the overlap section for vanilla and chocolate.

V is the total of the numbers in the circle for vanilla.

---

1 There are 148 students studying a language in Year 11 at Jamal's school.
Students study either French or Spanish, or both.
29 students study both French and Spanish.
93 students study French.

a Draw a Venn diagram to show this information.

Hint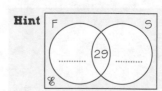

A student is selected at random.

b Work out the probability that the student studies Spanish. ....

c Given that the student studies Spanish, work out the probability that this student studies French. ....

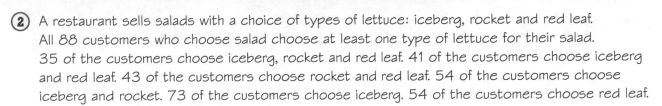

(2) A restaurant sells salads with a choice of types of lettuce: iceberg, rocket and red leaf.
All 88 customers who choose salad choose at least one type of lettuce for their salad.
35 of the customers choose iceberg, rocket and red leaf. 41 of the customers choose iceberg
and red leaf. 43 of the customers choose rocket and red leaf. 54 of the customers choose
iceberg and rocket. 73 of the customers choose iceberg. 54 of the customers choose red leaf.

a Draw a Venn diagram to show this information.

A customer is selected at random.

b Work out the probability that this customer chooses rocket.   ....

c Given that the customer chooses rocket, work out the probability that this
customer chooses iceberg.   ....

(3) Clare surveys 224 Year 11 students to see how many own smartphones, tablets and consoles.
29 don't own a smartphone, a tablet or a console. 22 of the students own a smartphone,
tablet and console. 50 of the students own a smartphone and console. 53 of the students own
a smartphone and tablet. 48 of the students own a tablet and console.
95 of the students own a smartphone. 100 of the students own a tablet.

a Draw a Venn diagram to show this information.

A student is selected at random.

b Given that the student owns a console, work out the probability that this
student owns a smartphone.   ....

Exam-style question

(4) There are 100 students studying science A-levels in a sixth form.
All 100 students study at least one area of science from physics, chemistry and biology.
9 of the students study physics, chemistry and biology. 34 of the students study physics and
chemistry. 19 of the students study chemistry and biology. 21 of the students study physics
and biology. 59 students study chemistry. 55 students study physics.

a Draw a Venn diagram to show this information.

(3 marks)

One of the 100 students is selected at random.

b Find the probability that this student studies chemistry but not physics. ....   (1 mark)

c Given that the student studies biology, find the probability that this
student also studies chemistry.   ....   (2 marks)

Reflect Do you always count all of the members in the sets to find a probability?
Explain your answer.

Practise the methods

Answer this question to check where to start.

Check up

There are nine counters in a bag. Four are green and five are red.
Two counters are picked at random. Tick the correct probability tree diagram.

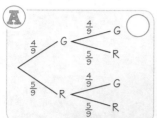
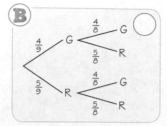
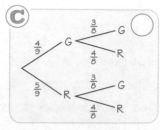
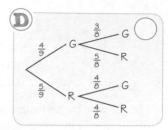

If you ticked D, work out the probability that both counters are the same colour. Then go to Q3.

If you ticked A, B or C, go to Q1 for more practice.

1 There are five counters in a bag. Two are green and three are red. Two counters are picked at random.

   a Work out the probability that the second counter is green, given that the first counter is green. ....

   b Work out the probability that the second counter is red, given that the first counter is red. ....

2 There are ten sweets in a bag. Six are orange and four are black. Two sweets are picked at random.

   a Work out the probability that the second sweet is orange, given that the first sweet is orange. ....

   b Work out the probability that the second sweet is orange, given that the first sweet is black. ....

3 A shop sells seven different flavours of ice cream with a choice of three toppings.
Work out the number of combinations of ice cream flavours and toppings. ....

Exam-style question

4 There are 15 yogurts on a tray. Seven are vanilla flavour and eight are toffee flavour.
Two yogurts are picked at random.
Work out the probability that both yogurts are the same flavour. .... **(4 marks)**

5 A pizza company offers a choice of three meat toppings: chicken, bacon and sausage.
264 of the pizzas are sold. 32 pizzas have chicken, bacon and sausage toppings.
53 pizzas have chicken and bacon. 117 pizzas have bacon and sausage.
36 pizzas have chicken and sausage. 135 pizzas have chicken. 145 pizzas have bacon.

   a Draw a Venn diagram to show this information.

One of the pizzas is selected at random.

   b Given that this pizza has a sausage topping, work out the probability that the pizza has a bacon topping. ....

Problem-solve!

1 Jon has 20 pieces of fruit in a fruit bowl.

He has 9 apples, 7 bananas and 4 plums.

Jon is going to take two pieces of fruit at random.

Work out the probability that the two pieces of fruit will not be the same type of fruit.

You must show all your working.

.... **(4 marks)**

2 There are ten counters in a box.

The letter A is on seven of the counters.

The letter B is on the other three counters.

Helen takes a counter from the box at random.

She keeps the counter.

Then Jane takes a counter from the box at random.

a Work out the probability that both Helen and Jane take a counter with the letter A on it.

.... **(3 marks)**

b Work out the probability that at least one counter with the letter A on it is taken.

.... **(2 marks)**

3 There are eight coins in a box.

Two coins are taken from the box at random.

Work out the probability that the total value of the two coins is at least 40p.

.... **(4 marks)**

4 Mikey has ten cards. There is a number on each card.

Mikey takes three of the cards at random.

He adds together the three numbers on the cards to get a total, T.

Work out the probability that T is an odd number.

.... **(4 marks)**

5 Laila has an empty box.
She puts some red counters and some blue counters into the box.
The ratio of the number of red counters to the number of blue counters is $1 : 9$.

Sam takes two counters from the box at random.

The probability that he takes two red counters is $\frac{1}{130}$.

How many red counters did Laila put into the box?

.... (4 marks)

6 There are ten chocolates in a box.
There are x milk chocolates in the box.
All the other chocolates are white chocolates.

Mia takes two chocolates from the box at random.

Find an expression, in terms of x, for the probability that
Mia takes one of each type of chocolate.

Give your answer in its simplest form.

.... (5 marks)

7 There are 250 members of a leisure centre.
The leisure centre has a gym, fitness classes and a swimming pool.
All 250 members use at least one of these facilities.

31 of the members use the gym, fitness classes and the swimming pool.
104 of the members use the gym and fitness classes.
45 of the members use the fitness classes and the swimming pool.
50 of the members use the gym and the swimming pool.
180 members use the gym.
97 members use the swimming pool.

a Draw a Venn diagram to show this information.

(3 marks)

One of the 250 members is selected at random.

b Find the probability that this member uses the gym but not the fitness classes.

.... (1 mark)

c Given that the member uses the fitness classes, find the probability
that this member also uses the gym.

.... (2 marks)

Now that you have completed this unit, how confident do you feel?

1 Combining events   **2 Tree diagrams**   **3 Venn diagrams for conditional probability**

⑥ Direct and inverse proportion

This unit will help you to solve problems where y is directly or inversely proportional to the square of x and to use and recognise graphs showing direct and inverse proportion.

A01 Fluency check

① **a** $y = 2x^2$

    i Work out y when $x = 5$. ....

    ii Work out x when $y = 128$. ....

    b $p = \dfrac{10}{q^2}$

    i Work out p when $q = 2$. ....

    ii Work out q when $p = 62.5$. ....

② Make x the subject of each equation.

  a $4y = x^2$ ....   **b** $y = 3x^2$ ....   **c** $y = \dfrac{5}{x^2}$ ....   **d** $ay = \dfrac{b}{x^2}$ ....

③ **Number sense**

Circle the odd one out.

    1   4   8   16   49   64

Key points

When y is directly proportional to the square of x, write $y \propto x^2$

When y is inversely proportional to the square of x, write $y \propto \dfrac{1}{x^2}$

These **skills boosts** will help you to solve problems involving direct and inverse proportion and to use and recognise graphs showing proportion.

| ① Direct proportion involving squares | ② Inverse proportion involving squares | ③ Direct and inverse proportion graphs |

You might have already done some work on proportion. Before starting the first skills boost, rate your confidence using these questions.

①

y is directly proportional to the square of x.

When $y = 32$, $x = 8$.

Work out x when $y = 200$.

②

y is inversely proportional to the square of x.

When $y = 0.16$, $x = 5$.

Work out y when $x = 100$.

③

In this graph, is y directly proportional to x? Explain your answer.

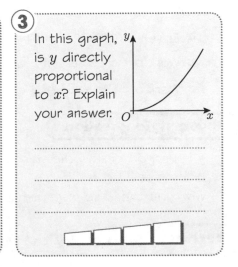

How confident are you?

1 Direct proportion involving squares

$y \propto x^2$ means 'y is directly proportional to the square of x'.
So $y = kx^2$, where k is a number/constant.

Guided practice

y is proportional to x^2. When $x = 5$, $y = 75$.

a Find a formula for y in terms of x.

b Work out the value of y when $x = 8$.

_____

a $y \propto x^2$

So $y = kx^2$

$75 = k \times \text{..........}^2$

$k = \text{..........}$

The formula is $y = \text{..........} x^2$

b When $x = 8$, $y = \text{..........}$

Substitute $x = 5$ and $y = 75$ into the equation $y = kx^2$

Simplify and rearrange to work out the value of k.

Substitute your value of k into the equation $y = kx^2$

Substitute $x = 8$ into your formula to work out y.

1 y is proportional to the square of x. When $x = 3$, $y = 45$.

  a Find a formula for y in terms of x. ....

  b Work out the value of y when $x = 6$. ....

2 y is proportional to x^2. When $x = 2$, $y = 16$.

  a Find a formula for y in terms of x. ....

  b Work out the value of y when $x = 7$. ....

  c Work out the value of x when $y = 100$. ....

Hint To find x when $y = 100$, substitute $y = 100$ into your formula and rearrange it to find x.

3 q is proportional to the square of p. When $p = 10$, $q = 20$.

  a Find a formula for q in terms of p. ....

  b Work out the value of q when $p = 4$. ....

  c Work out the value of p when $q = 0.05$. ....

4 b is proportional to a^2. When $a = 4$, $b = 8$.

  a Find a formula for b in terms of a. ....

  b Work out the value of b when $a = 12$. ....

  c Work out the value of a when $b = 0.98$. ....

Exam-style question

5 y is proportional to the square of x.
When $x = 10$, $y = 25$.
Work out the value of y when $x = 8.6$. .... **(3 marks)**

Reflect Explain what $y \propto x^2$ means and then use what you know to explain what $y \propto \sqrt{x}$ means.

2 Inverse proportion involving squares

$y \propto \dfrac{1}{x^2}$ means 'y is inversely proportional to the square of x'.

So $y = \dfrac{k}{x^2}$ where k is a number/constant.

Guided practice

y is inversely proportional to the square of x.

When $x = 10$, $y = 0.04$.

a Find a formula for y in terms of x.

b Work out the value of y when $x = 5$.

Worked exam question

a $y \propto \dfrac{1}{x^2}$

So $y = \dfrac{k}{x^2}$

$\ldots\ldots = \dfrac{k}{10^2}$

$k = \ldots\ldots$

The formula is $y = \dfrac{\ldots\ldots}{x^2}$

b When $x = 5$, $y = \ldots\ldots$

Substitute, $x = 10$ and $y = 0.04$ into the equation $y = \dfrac{k}{x^2}$

Simplify and rearrange to work out the value of k.

Substitute your value of k into the equation $y = \dfrac{k}{x^2}$

Substitute $x = 5$ into your formula to work out y.

① y is inversely proportional to the square of x. When $x = 0.5$, $y = 20$.

    a Find a formula for y in terms of x.   ....

    b Work out the value of y when $x = 2$.   ....

② y is inversely proportional to x^2. When $x = 2$, $y = 0.125$.

    a Find a formula for y in terms of x.   ....

    b Work out the value of y when $x = 0.4$.   ....

    c Work out the value of x when $y = 1250$.   ....

Hint To find x when $y = 1250$, substitute $y = 1250$ into your formula and rearrange it to find x.

③ q is inversely proportional to the square of p. When $p = 0.1$, $q = 80$.

    a Find a formula for q in terms of p.   ....

    b Work out the value of q when $p = 5$.   ....

    c Work out the value of p when $q = 5$.   ....

Exam-style question

④ y is inversely proportional to the square of x.

When $x = 0.2$, $y = 6.25$.

Work out the value of y when $x = 0.1$.   .... **(3 marks)**

Reflect

Explain what $y \propto \dfrac{1}{x^2}$ means and then use what you know to explain what $y \propto \dfrac{1}{x^3}$ means.

3 Direct and inverse proportion graphs

In this graph, y is directly proportional to x.

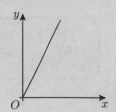

In this graph, y is directly proportional to the square of x.

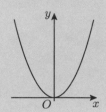

In this graph, y is inversely proportional to x.

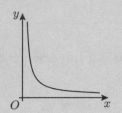

In this graph, y is inversely proportional to the square of x.

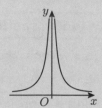

Guided practice

In which of these graphs is y directly proportional to x?

A 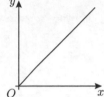   **B** 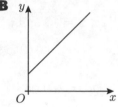   **C** 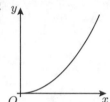   **D**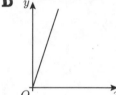

In graphs A and .... y is directly proportional to x.

If y is directly proportional to x, the graph will be a straight line that passes through the origin. Graph B does not pass through the origin.

1 In this graph y is directly proportional to x. Find the values of a and b.

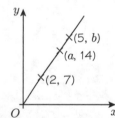

(5, b)
(a, 14)
(2, 7)

2 In this graph y is inversely proportional to x. Find the values of p and q.

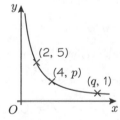

(2, 5)
(4, p) (q, 1)

....

Exam-style question

3 Each graph shows a proportionality relationship between y and x.

Match each graph with a statement in the table below.

Graph A 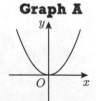   **Graph B**   **Graph C** 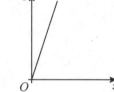   **Graph D**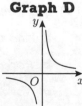

Proportionality statement	Graph
y is directly proportional to x.	....
y is inversely proportional to x.	....
y is directly proportional to the square of x.	....
y is inversely proportional to the square of x.	....

(2 marks)

Reflect Describe the features of the graphs for direct and inverse proportion.

Practise the methods

Answer this question to check where to start.

Check up

Tick the equation that represents the relationship 'y is inversely proportional to the square of x'.

 A $y = kx$ ◯

 B $y = kx^2$ ◯

 C $y = \dfrac{k}{x^2}$ ◯

 D $y = \dfrac{x^2}{k}$ ◯

If you ticked C, work out the value of k given that $x = 2$ when $y = 2.5$. Then go to Q4.

If you ticked A, B or D go to Q1 for more practice.

---

① Match each proportionality statement with an equation.

A y is directly proportional to x.

B y is inversely proportional to x.

C y is directly proportional to the square of x.

D y is inversely proportional to the square of x.

i $y = \dfrac{k}{x}$

ii $y = kx^2$

iii $y = \dfrac{k}{x^2}$

iv $y = kx$

---

② y is directly proportional to the square of x.

When $x = 1.5$, $y = 18$.

Find a formula for y in terms of x. ....

③ y is inversely proportional to the square of x.

When $x = 0.2$, $y = 250$.

Find a formula for y in terms of x. ....

④ y is directly proportional to the square of x.

When $x = 4$, $y = 20$.

a Find a formula for y in terms of x. ....

b Work out the value of y when $x = 18$. ....

⑤ Circle the graph in which y is inversely proportional to x.

A 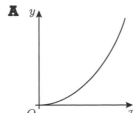   **B** 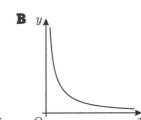   **C**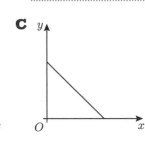

Exam-style question

⑥ T is inversely proportional to d^2.

$T = 180$ when $d = 6$.

Find the value of T when $d = 0.5$. .... **(3 marks)**

Problem-solve!

Exam-style question

(1) D is directly proportional to the square of n.

Carly says that when n is doubled, the value of D is doubled.

Carly is wrong.

Explain why. .... **(1 mark)**

(2) In an experiment, measurements of a and b were taken.

a	2	6	10
b	4	108	500

Which of these relationships fits the result?

$b \propto a$ $b \propto a^2$ $b \propto a^3$ $b \propto \sqrt{a}$ ....

Exam-style questions

(3) y is inversely proportional to the square of x.

When $x = 4$, $y = 1440$.

Work out the value of y when $x = 0.25$.   .... **(3 marks)**

(4) y is directly proportional to the square of x.

When $x = 20$, $y = 240$.

Work out the value of y when $x = 3$.   .... **(3 marks)**

(5) d is inversely proportional to the square of c.

When $c = 5$, $d = 300$.

Work out the value of d when $c = 2.5$.   .... **(3 marks)**

(6) In this graph y is inversely proportional to x.

Find the values of a and b.

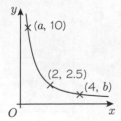

....

Now that you have completed this unit, how confident do you feel?

1 Direct proportion involving squares	**2** Inverse proportion involving squares	**3** Direct and inverse proportion graphs

(7) Accuracy and bounds

This unit will help you to calculate using upper and lower bounds.

AO1 Fluency check

1. Round each number to the level of accuracy given.

    a 3.5 to the nearest whole number ....   **b** 5.844 to 2 decimal places (d.p.) ....

    c 9.99 to 1 significant figure (s.f.) ....   **d** 23 466 to 2 s.f. ....

    e 24.666 to 3 s.f. ....   **f** 5.555 to 1 d.p. ....

2. For each measurement, write **i** the upper bound (u.b.) **ii** the lower bound (l.b.).

    a 12 cm to the nearest cm   **b** 5.23 km to 2 d.p.   **c** 7000 kg to 1 s.f.

    i .... **ii** ....   **i** .... **ii** ....   **i** .... **ii** ....

    d 3.6 litres to 1 d.p.   **e** 8500 m to 2 s.f.   **f** 6.4 km to the nearest tenth of a km

    i .... **ii** ....   **i** .... **ii** ....   **i** .... **ii** ....

3. **Number sense**

    Work out

    a 1.5 × 0.9 ....   **b** 8.5 × 0.5 ....   **c** 2.5 ÷ 0.25 ....   **d** 7.6 ÷ 0.2 ....

Key points

The upper and lower bounds are the largest and smallest possible values, given the rounded measurements you are told.

For a length, l m, given as 21.7 m to 1 d.p., the upper bound is 21.75 m and the lower bound is 21.65 m.

The inequality to show this is $21.65 \leqslant l < 21.75$.

These **skills boosts** will help you to use upper and lower bounds in calculations.

1 Upper and lower bounds of measurements

2 Upper and lower bounds of compound measures

You might have already done some work on accuracy and bounds. Before starting the first skills boost, rate your confidence using these questions.

1 The length of a rectangle is 7.2 cm and its width is 4.6 cm, both to the nearest mm. Calculate the upper and lower bounds for the area of the rectangle.

2 The mass of an iron bar is 650 g to 2 s.f. The volume of the bar is 82.6 cm³ to 1 d.p. Calculate the upper and lower bounds for the density of iron.

How confident are you?

1 Upper and lower bounds of measurements

To find the upper bound (u.b.) for calculations using addition or multiplication, use the u.b. of the measurements.

Guided practice

A rectangle measures 7.9 cm by 12.4 cm to 1 d.p.
Calculate the upper and lower bounds for the area of the rectangle.

Width = 7.9 cm (to 1 d.p.)

Upper bound = ... cm, lower bound = 7.85 cm

Length = 12.4 cm (to 1 d.p.)

Upper bound = 12.45 cm, lower bound = ... cm

Area = length × width

Upper bound = ... × .... = ... cm²

Lower bound = ... × .... = ... cm²

The u.b. has a 5 in the second decimal place.
To find the l.b., subtract 1 from the digit in the first decimal place and put a 5 in the second decimal place.

Upper bound for the area: multiply the u.b. of the length by the u.b. of the width.
Lower bound for the area: multiply the l.b. of the length by the l.b. of the width.
Do not round your answers.

(1) A rectangle measures 2.7 m by 4.0 m to 1 d.p.
Calculate the upper and lower bounds for the area of the rectangle. ....

(2) Calculate the upper and lower bounds for the perimeter, P, of each shape.
All measurements are rounded to 1 d.p.
Give your answers as inequalities.

Hint A number rounded to 1 d.p. always has 1 d.p.
The digit in that decimal place might be 0.

a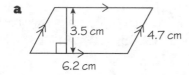

b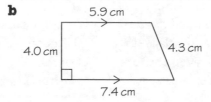

....

(3) Calculate the upper and lower bound for the circumference of the circle.
All measurements are rounded to 1 d.p.
Give your answers as an inequality in terms of π.

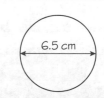

....

(4) Calculate the upper and lower bounds for the area, A, of each shape in Q2.
Give your answers as inequalities. ....

Exam-style question

(5) Triangle ABC is a right-angled triangle.
AB is 3.7 cm to 1 d.p.
AC is 2.9 cm to 1 d.p.
Calculate the upper bound for length BC, giving your answer
to 2 d.p.

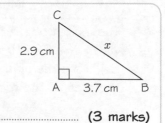

.... **(3 marks)**

Reflect How did you work out the upper and lower bounds for your calculations?

2 Upper and lower bounds of compound measures

To find the upper bound (u.b.) for calculations involving division or subtraction, use the u.b. and subtract or divide by the lower bound (l.b.).

To find the l.b. involving division or subtraction, use the l.b. and subtract or divide by the u.b.

Guided practice

Worked exam question

An athlete runs 100 m, to the nearest metre, in 11.2 seconds, to 1 d.p. Work out the upper and lower bounds for the speed of the athlete.

Distance = 100 m (to the nearest metre)

First find the upper and lower bounds for the distance and the time.

Upper bound = ... m, lower bound = ... m

Time = 11.2 s (to 1 d.p.)

Upper bound = ... s, lower bound = ... s

$$Speed = \frac{distance}{time}$$

$$Upper\ bound\ for\ speed = \frac{distance\ upper\ bound}{time\ lower\ bound}$$

= .... ÷ .... = ... m/s

$$Lower\ bound\ for\ speed = \frac{distance\ lower\ bound}{time\ upper\ bound}$$

= .... ÷ .... = ... m/s

When dividing, the u.b. is achieved by dividing the largest number possible by the smallest number possible.

To get the l.b., you divide the smallest number possible by the largest number possible. If your answer is not exact and you are not told how to round it, give it to two more decimal places than the maximum number of decimal places in any of the measurements. In this case, round your answers to 3 d.p.

① A car travels 180 miles, to the nearest mile, at an average speed of 52.4 mph, to 1 d.p.

Calculate the upper and lower bounds for the time the journey takes. ....

② A bangle has a mass of 17.4 g and a volume of 0.9 cm³, both measured to 1 d.p.

Hint $density = \frac{mass}{volume}$

Calculate the upper and lower bounds for the density of the bangle. ....

③ A force of 42 N, measured to the nearest whole number, is applied to an area of 2.3 m², measured to 1 d.p.

Hint $pressure = \frac{force}{area}$

Calculate the upper and lower bounds for the pressure in N/m². ....

Exam-style question

④ A solid sphere has a mass of 1250 g, measured to the nearest 10 g, and a volume of 173 cm³, measured to the nearest cm³.

Given that $density = \frac{mass}{volume}$

work out the upper bound for the density of the sphere.

Give your answer to 3 s.f. .... (3 marks)

Reflect

How did you work out the upper and lower bounds for your calculations?

Practise the methods

Answer this question to check where to start.

Check up

The mass of a steel bar is 480 g to the nearest 10 grams. The volume of the bar is 61.1 cm³ to 1 d.p. Tick the correct calculation for the upper bound of the density.

 **A** $\dfrac{485}{61.15}$ ○

 **B** $\dfrac{475}{61.05}$ ○

 **C** $\dfrac{485}{61.05}$ ○

 **D** $\dfrac{475}{61.15}$ ○

If you ticked C, calculate the upper bound of the density of the steel bar. Then go to Q3.

If you ticked A, B or D, go to Q1 for more practice.

① $R = \dfrac{p}{q}$

a Work out the upper and lower bounds of R when $p = 7$ to the nearest whole number and $q = 1.5$ to 1 d.p. ....

b Work out the upper and lower bounds of R when $p = 260$ to the nearest 10 and $q = 12.3$ to 1 d.p. ....

c Work out the upper and lower bounds of R when $p = 800$ to 1 s.f. and $q = 18$ to the nearest whole number. ....

② $C = \dfrac{a}{b}$

a Work out the upper and lower bounds of C when $a = 25$ to the nearest whole number and $b = 3.75$ to 2 d.p. ....

b Work out the upper and lower bounds of C when $a = 3200$ to 2 s.f. and $b = 34.7$ to 3 s.f. ....

c Work out the upper and lower bounds of C when $a = 500$ to 1 s.f. and $b = 9.2$ to 1 d.p. ....

③ A force of 35 N, measured to the nearest whole number, is applied to an area of 1.8 m², measured to 1 d.p.

Calculate the upper and lower bounds for the pressure in N/m². ....

④ A rectangle has a diagonal of 8.2 cm measured to 1 d.p. and a length of 6 cm measured to the nearest cm.
Work out the upper bound of the width of the rectangle.

....

Exam-style question

⑤ A solid has a mass of 900 g, measured to 1 s.f. and a volume of 800 cm³, measured to the nearest cubic centimetre.

Given that $\text{density} = \dfrac{\text{mass}}{\text{volume}}$

work out the lower bound for the density of the solid.

Give your answer to 3 s.f. .... **(3 marks)**

Problem-solve!

1 $I = \dfrac{V}{R}$

$V = 220$ correct to the nearest 5.

$R = 3500$ correct to the nearest 100.

Work out the lower bound for the value of I.

Give your answer correct to 3 d.p.

You must show your working. .... (3 marks)

2 Hamza travelled from Swindon to Manchester.

He travelled 165 miles, correct to the nearest 5 miles.

The journey took him 200 minutes, correct to the nearest 10 minutes.

Calculate the lower bound for the average speed of the journey.

Give your answer in **miles per hour**, correct to 3 s.f.

You must show all your working. .... (4 marks)

3 A solid sphere has a mass of 515 g, measured to the nearest gram,

and a radius of 4.7 cm, measured to the nearest millimetre.

Given that   density $= \dfrac{\text{mass}}{\text{volume}}$

find the upper bound for the density of the sphere.

Give your answer to 3 s.f. .... (4 marks)

4 Olivia drops a ball from a height of d metres to the ground.

The time, t seconds, that the ball takes to reach the ground is given by $t = \sqrt{\dfrac{2d}{g}}$,

where g m/s^2 is the acceleration due to gravity.

Olivia measures d as 24.7 m correct to 3 s.f.

$g = 9.8$ m/s^2 correct to 2 s.f.

Calculate the lower bound of t.

Give your answer to 3 s.f.

You must show all your working. .... (4 marks)

5 Jayden wants to cover a triangular field, ABC, with fertiliser.

Here are the measurements Jayden makes.

Angle ABC $= 55°$ correct to the nearest degree.

BA $= 175$ m correct to the nearest 5 m.

BC $= 100$ m correct to the nearest 5 m.

Work out the upper bound for the area of the field.

Give your answer to the nearest m^2.

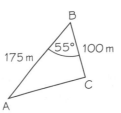

You must show your working. .... (3 marks)

Now that you have completed this unit, how confident do you feel?

1 Upper and lower bounds of measurements

2 Upper and lower bounds of compound measures

Answers

Unit 1 Statistics

1. **a** A population is the whole set of data.
   b A census is a survey of the whole population.
   c A sample is part of a population, usually at least 10%.
2. Draw names from a hat or use a table of random numbers.

3 Number sense

a 25   **b** 13   **c** 8.5   **d** 12.4

Confidence questions

1. The best sampling method would be stratified sampling because the data is grouped into year groups. In a stratified sample the number of students from each year group will be proportional to the number of students in that year group in the population so the sample will be more representative.
2. 5.3   **3** 8

Skills boost 1 Sampling

Guided practice

10%

The best sampling method would be stratified sampling because the data is grouped into year groups. In a stratified sample the number of students from each year group will be proportional to the number of students in that year group in the population so the sample will be more representative.

1. Random sample – there are no groups specified so the sampling method should give every customer an equal chance of being selected.
2. Stratified sample – the data is grouped by age. In a stratified sample the number of people from each age group will be proportional to the number of people in that age group in the population so the sample will be more representative.
3. **a** Draw names from a hat or use a table of random numbers.
   b Every item has an equal chance of being chosen.

Skills boost 2 Drawing histograms

Guided practice

Class width: 20 (given), 20, 20, 20, 20
Frequency density: 22.8 (given), 56.15, 41.4, 16.6, 4.35

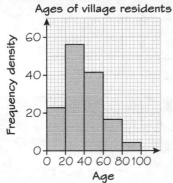

Ages of village residents

1. Class width: 5 for all groups
   Frequency density: 0.4, 1.6, 2.2, 1.2, 1

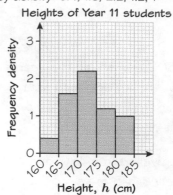

Heights of Year 11 students

2. Class width: 5 for all groups
   Frequency density: 4, 6, 11, 12, 1

Times taken to travel to school on Monday

3. Frequency: 6, 9, 23, 25, 2
4. **a** $800 < d \leqslant 1000$ **b** 20   **c** 525

Skills boost 3 Estimating the mean from a histogram

Guided practice

Frequency: 19 (given), 15, 9, 12, 2
Midpoint: 2 (given), 6, 10, 14, 18
Frequency × Midpoint: 38 (given), 90, 90, 168, 36
Estimated total height of all the trees = 422
Total number of trees = 57
Estimate of mean = 422 ÷ 57 = 7.4 metres

1. 31 minutes   **2** 59.2 mm
2. 14.4 minutes   **4** 3.22 kg

Practise the methods

1. **a** Frequency density: 0.7, 1.15, 0.75, 0.4
   b Frequency density: 0.12, 0.47, 0.36, 0.14

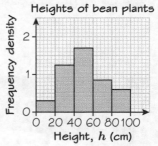

Heights of bean plants

2. **a** $5 < d \leqslant 10$ **b** 21   **c** 50

Problem-solve!

1. Stratified sample because the data is grouped by age. In a stratified sample the number of people from each age group will be proportional to the number of people in that age group in the population so the sample will be more representative.

2. Frequency density: 10.3, 11.4, 6.5, 0.2

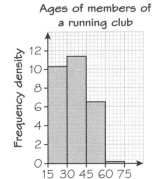

Ages of members of a running club

3. **a** Frequency: 9
   b Frequency density: 0.2 (given), 0.27, 0.67, 0.3 (given), 0.17

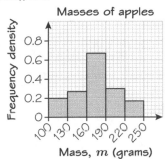

Masses of apples

   c 174.4 g

Unit 2 Indices and recurring decimals

AO1 Fluency check

1. **a** $7n$ **b** $3n$ **c** $11n$ **d** n
2. **a** 64 **b** $\frac{1}{5}$ **c** $\frac{1}{16}$ **d** 8
3. **a** 32 **b** 125 **c** 729 **d** $100\,000\,000$

4 Number sense

   a 33.3 **b** 521.21 **c** $63\,591.591$

Confidence questions

1. $\frac{5}{11}$ 2. $\frac{1}{9}$

Skills boost 1 Converting recurring decimals to fractions

Guided practice

$$10x - x = 4$$
$$9x = 4$$
$$x = \frac{4}{9}$$

1. **a** $\frac{7}{9}$ **b** $\frac{2}{9}$ **c** $\frac{13}{99}$ **d** $\frac{6}{11}$
   e $\frac{95}{333}$ **f** $\frac{28}{37}$

2. **a** $1\frac{1}{3}$ **b** $4\frac{26}{33}$ **c** $6\frac{5}{37}$ **d** $\frac{23}{45}$
   e $\frac{1}{6}$ **f** $\frac{241}{495}$

3. $\frac{7}{18}$

Skills boost 2 Calculating using fractional and negative indices

Guided practice

a $64^{\frac{1}{3}} = \sqrt[3]{64} = 4$

b $(10^2)^{\frac{1}{2}} = 10^{2 \times \frac{1}{2}} = 10^1 = 10$

c $32^{-\frac{2}{5}} = \frac{1}{32^{\frac{2}{5}}} = \frac{1}{(\sqrt[5]{32})^2} = \frac{1}{2^2} = \frac{1}{4}$

1. **a** 5   **b** 9   **c** 12   **d** 2
   e 10   **f** 15   **g** 5   **h** 13
2. **a** 2   **b** 10   **c** 3   **d** 2
3. **a** $\frac{1}{10}$ **b** $\frac{3}{4}$ **c** $\frac{5}{8}$ **d** $\frac{3}{10}$
4. **a** $\frac{1}{2}$ **b** $\frac{1}{3}$ **c** $\frac{1}{2}$ **d** $\frac{1}{7}$
5. **a** 4   **b** 7   **c** 3   **d** 10
6. **a** 2   **b** 5   **c** 3   **d** 4
7. **a** 5   **b** 10   **c** 7   **d** 9
8. **a** 2   **b** $\frac{1}{4}$ **c** 3   **d** $\frac{1}{5}$
9. **a** 9   **b** 512   **c** 16   **d** 25
   e 27   **f** 8   **g** 216   **h** 100
10. **a** $\frac{1}{4}$ **b** $\frac{1}{5}$ **c** $\frac{1}{3}$ **d** $\frac{1}{11}$
11. **a** $\frac{1}{4}$ **b** $\frac{1}{16}$ **c** $\frac{1}{27}$ **d** $\frac{1}{27}$
    e $\frac{1}{100}$ **f** $\frac{1}{8}$ **g** $\frac{1}{49}$ **h** $\frac{1}{8}$
12. **a** 4   **b** 5   **c** $\frac{1}{9}$

Practise the methods

1. **a** $\frac{3}{10}$ **b** $\frac{7}{10}$ **c** $\frac{33}{100}$ **d** $\frac{373}{1000}$
2. **a** $\frac{1}{9}$ **b** $\frac{8}{33}$ **c** $\frac{401}{999}$
3. $\frac{13}{18}$
4. **a** 6   **b** 5   **c** 3   **d** 10
5. **a** 6   **b** 5   **c** 3   **d** 10
6. They are the same.
7. $16^{\frac{1}{2}} = \pm4$, $16^{-\frac{1}{2}} = \pm\frac{1}{4}$, $4^{\frac{3}{2}} = \pm8$, $4^{-\frac{3}{2}} = \pm\frac{1}{8}$, $64^{\frac{2}{3}} = 16$, $64^{-\frac{2}{3}} = \frac{1}{16}$
8. **a** 10   **b** $\frac{16}{25}$ **c** $x = 4$

Problem-solve!

1. $\frac{16}{75}$
2. **a** 2   **b** 4   **c** $\frac{125}{64}$
3. **a** 3   **b** $\frac{1}{7}$ **c** $\frac{1}{4}$
4. **a** x **b** 2
5. -5, $25^{-\frac{1}{2}}$, 0.5, $25^{\frac{1}{2}}$

Unit 3 Surds

AO1 Fluency check

① **a** $x^2 - x - 6$ **b** $x^2 - 25$
   c $2x^2 - 15x + 28$ **d** $2x^2 - x - 15$

② **a** $\dfrac{1}{2}$ **b** $\dfrac{1}{2}$ **c** $\dfrac{7}{9}$ **d** $\dfrac{3}{8}$

③ **a** 25 **b** 4 **c** 7 **d** 3

④ Number sense

   a 5×9 **b** 3×25
   c 5×16 or 4×20 **d** 3×49

Confidence questions

① $2\sqrt{3}$

② $7 - 4\sqrt{3}$

③ $\dfrac{2\sqrt{5}}{5}$

④ $1 - 2\sqrt{x} + x$

Skills boost 1 Simplifying surds

Guided practice

$\sqrt{12} = \sqrt{4 \times 3} = \sqrt{4} \times \sqrt{3} = 2 \times \sqrt{3} = 2\sqrt{3}$

① **a** $3\sqrt{5}$ **b** $5\sqrt{3}$ **c** $4\sqrt{3}$ **d** $4\sqrt{5}$
   e $10\sqrt{2}$ **f** $7\sqrt{3}$ **g** $5\sqrt{2}$ **h** $4\sqrt{2}$

② **a** $4\sqrt{3}$ **b** $6\sqrt{7}$ **c** $15\sqrt{3}$ **d** $6\sqrt{2}$

③ **a** $\dfrac{\sqrt{7}}{3}$ **b** $\dfrac{\sqrt{11}}{5}$ **c** $\dfrac{2\sqrt{3}}{5}$ **d** $\dfrac{2\sqrt{7}}{7}$

④ $2\sqrt{5}$

Skills boost 2 Expanding brackets involving surds

Guided practice

$(2 - \sqrt{3})(1 + \sqrt{3}) = 2(1 + \sqrt{3}) - \sqrt{3}(1 + \sqrt{3})$
$= 2 + 2\sqrt{3} - \sqrt{3} - 3$
$= -1 + \sqrt{3}$

① **a** -2 **b** $1 + \sqrt{5}$ **c** $5 + 4\sqrt{2}$
   d $13 + 3\sqrt{11}$ **e** $11 - 5\sqrt{7}$ **f** $-2 - 2\sqrt{5}$

② **a** $3 + 2\sqrt{2}$ **b** $9 - 4\sqrt{5}$ **c** $16 + 6\sqrt{7}$
   d $4 - 2\sqrt{3}$

③ $12 - 6\sqrt{3}$

④ 27

⑤ $1 - 2\sqrt{2}$

Skills boost 3 Rationalising denominators

Guided practice

$\dfrac{4}{\sqrt{5}} = \dfrac{4}{\sqrt{5}} \times \dfrac{\sqrt{5}}{\sqrt{5}} = \dfrac{4 \times \sqrt{5}}{\sqrt{5} \times \sqrt{5}} = \dfrac{4\sqrt{5}}{5}$

① **a** $\dfrac{\sqrt{2}}{2}$ **b** $\dfrac{\sqrt{7}}{7}$ **c** $\dfrac{\sqrt{3}}{3}$ **d** $\dfrac{\sqrt{5}}{5}$
   e $\dfrac{\sqrt{11}}{11}$ **f** $\dfrac{\sqrt{13}}{13}$ **g** $\dfrac{\sqrt{17}}{17}$ **h** $\dfrac{\sqrt{19}}{19}$

② **a** $\dfrac{3\sqrt{2}}{2}$ **b** $\dfrac{2\sqrt{3}}{3}$ **c** $\dfrac{3\sqrt{5}}{5}$ **d** $\dfrac{4\sqrt{3}}{3}$
   e $\dfrac{5\sqrt{7}}{7}$ **f** $\dfrac{2\sqrt{5}}{5}$ **g** $\dfrac{4\sqrt{7}}{7}$ **h** $\dfrac{6\sqrt{11}}{11}$

③ **a** $2\sqrt{2}$ **b** $\sqrt{3}$ **c** $\sqrt{2}$ **d** $2\sqrt{3}$
   e $2\sqrt{7}$ **f** $\sqrt{5}$ **g** $\sqrt{7}$ **h** $2\sqrt{11}$

④ $2\sqrt{5}$

Skills boost 4 Simplifying algebraic expressions involving surds

Guided practice

a $(x + \sqrt{y})^2 = x(x + \sqrt{y}) + \sqrt{y}(x + \sqrt{y})$
        $= x^2 + x\sqrt{y} + x\sqrt{y} + y$
        $= x^2 + 2x\sqrt{y} + y$

b $\dfrac{1}{\sqrt{x}} = \dfrac{1}{\sqrt{x}} \times \dfrac{\sqrt{x}}{\sqrt{x}} = \dfrac{\sqrt{x}}{x}$

① **a** $a^2 - b$ **b** $m^2 - 2m\sqrt{n} + n$
   c $p^2 - q$

② **a** $2a^2 - a\sqrt{b} - 6b$ **b** $j^2 + 8j\sqrt{k} + 16k$
   c $2c^2 - 3c\sqrt{d} - 2d$

③ **a** $\dfrac{\sqrt{a}}{a}$ **b** $\dfrac{2\sqrt{p}}{p}$ **c** $\dfrac{5\sqrt{t}}{t}$

④ **a** $\dfrac{\sqrt{2b}}{2b}$ **b** $\dfrac{2\sqrt{3u}}{u}$ **c** $\dfrac{\sqrt{5e}}{e}$

⑤ $a - 9b$

Practise the methods

Check up $10\sqrt{3}$

① **a** $\sqrt{64 \times 2}$ **b** $8\sqrt{2}$

② **a** $5\sqrt{5}$ **b** $9\sqrt{2}$ **c** $12\sqrt{2}$ **d** $15\sqrt{2}$

③ **a** $\dfrac{3\sqrt{5}}{5}$ **b** $4\sqrt{3}$ **c** $5\sqrt{2}$ **d** $\sqrt{k}$

④ **a** -4 **b** $17 + 5\sqrt{11}$ **c** $4a^2 + 4a\sqrt{b} + b$

⑤ $10\sqrt{3}$

⑥ $30 + 10\sqrt{5}$

Problem-solve!

① $3\sqrt{7}$

② $12 - 6\sqrt{3}$

③ $3\sqrt{3}$

④ $\dfrac{(1 + \sqrt{3})}{\sqrt{2}} \times \dfrac{\sqrt{2}}{\sqrt{2}} = \dfrac{\sqrt{2}(1 + \sqrt{3})}{2} = \dfrac{\sqrt{2} + \sqrt{6}}{2}$

⑤ $12\sqrt{2}$

⑥ $x - 4y$

⑦ $3.75\,\text{cm}$

Unit 4 Cumulative frequency

AO1 Fluency check

① 9 ② 34.5 ③ 5

④ Number sense

   a 213 **b** 304

Confidence questions

① Cumulative frequency: 5, 20, 43, 55, 58

② The median of the data set

③ The boys' results were more spread out than the girls'.

Skills boost 1 Constructing cumulative frequency graphs

Guided practice

$0 < h \leqslant 30$: cumulative frequency $= 3 + 21 + 35$
$= 59$; 59 plants are up to 30 cm tall

$0 < h \leqslant 40$: cumulative frequency $= 3 + 21 + 35$
$+ 16 = 75$; 75 plants are up to 40 cm tall

$0 < h \leqslant 50$: cumulative frequency $= 3 + 21 + 35$
$+ 16 + 5 = 80$; 80 plants are up to 50 cm tall

Heights of bean plants

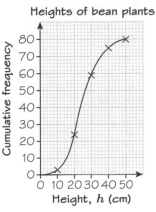

(1) Cumulative frequency: 21, 53, 75, 92, 100

Masses of parcels

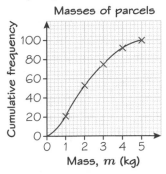

(2) Cumulative frequency: 3, 15, 34, 55, 60

Times taken to get to work

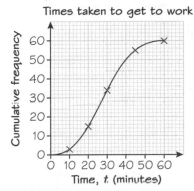

Skills boost 2 Drawing and interpreting box plots

Guided practice

a Upper quartile = 31

Ages of women in a running club

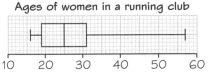

b Interquartile range (IQR) = 31 − 19 = 12
1.5 × IQR = 18
Lower quartile − 1.5 × IQR = 19 − 18 = 1
Upper quartile + 1.5 × IQR = 31 + 18 = 49
Outliers: 57

(1) **a** Numbers of points scored   **b** 6

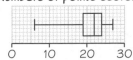

(2) **a**   **Weekly wages of staff**

100   200   300   400   500   600   700

b There are no outliers.

---

(3)   **Revision times**

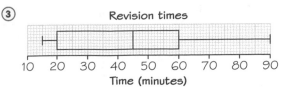

(4)   **Heights of sunflowers**

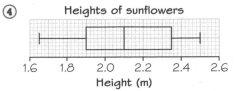

(5) **a** 15.5°C
   b Maximum = 29°C and minimum = 3°C
   c 10°C

(6) **a** 47 kg   **b** 10 kg   **c** 25 children

Skills boost 3 Comparing distributions

Guided practice

a Median = 13°C
b Interquartile range = upper quartile − lower quartile
            = 14 − 10 = 4°C
c Median for Glasgow = 10.5°C
   Interquartile range for Glasgow = upper quartile −
                         lower quartile
            = 13 − 8 = 5°C

The median temperature for London is 13 °C and
for Glasgow is 10.5 °C, so average temperature for
London is higher.
The interquartile range for Glasgow is slightly greater
than for London suggesting that the temperatures
in Glasgow are slightly more spread out than the
temperatures in London.

(1) **a** 24 mg
   b 50 mg
   c Median amount of vitamin C in vegetables = 17 mg
      Interquartile range of amounts of vitamin C in
      vegetables = 33 mg
      On average, the fruits have more vitamin C than
      the vegetables. The interquartile range for the
      fruits is greater than the interquartile range for
      the vegetables which suggests that the amounts
      of vitamin C in fruit is more spread out than the
      amounts of vitamin C in vegetables.

(2) Median length of call for employee A = 13 minutes
   Median length of call for employee B = 28 minutes
   Interquartile range for employee A = 12.5 minutes
   Interquartile range for employee B = 21 minutes
   On average, employee A's calls are shorter than
   employee B's calls.
   The lengths of employee A's calls are less spread out
   than the lengths of employee B's calls.

(3) **a** 5 minutes   **b** 9.5 minutes
   c Median waiting time for afternoon clinic = 8 minutes
      Interquartile range for afternoon clinic = 8 minutes
      On average, the waiting times were longer in the
      afternoon.
      The waiting times were more spread out in the
      morning.

Practise the methods

Check up Median = 88 cm; interquartile range = 14 cm

① **a** (50, 3), (60, 10), (70, 26), (80, 36), (90, 40)

   b (0.5, 3), (1, 9), (1.5, 23), (2, 28), (2.5, 30)

② **a**

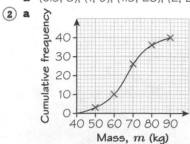

Mass, *m* (kg)

   b

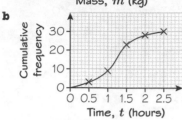

Time, *t* (hours)

③ The median result for group A is 62 and for group B is 74, so on average the test results for group B are higher.

The interquartile range for group A is 14 and for group B is 10, so the results for group B are more consistent/less spread out.

Problem-solve!

① **a** Cumulative frequency: 3, 18, 53, 79, 94, 100

   b Times taken in a fun run

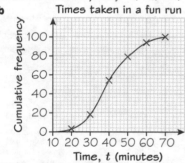

Time, *t* (minutes)

   c 39 minutes

   d 16 people

②

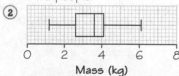

Mass (kg)

③ The median heart rate before walking up the stairs is 68 beats per minute and the median after is 78 beats per minute.

So on average the heart rates are higher after walking up the stairs.

The interquartile range before walking up the stairs is 14 beats per minute and the interquartile range after is 23 beats per minute.

So the heart rates after walking up the stairs are more spread out than the heart rates before walking up the stairs.

④ The median for group A is 161.25 cm and the median for group B is 167.5 cm.

So on average, the children in group B are taller.

The interquartile range for group A is 5.25 cm and for group B is 4.75, so the heights of the children in group A and group B are similarly spread out.

Unit 5 Probability

AO1 Fluency check

① 0.6   ② $\frac{1}{24}$

③ **Number sense**

   a 0.14   **b** 0.03   **c** $\frac{2}{15}$ **d** $\frac{5}{8}$

Confidence questions

① 15   ② $\frac{5}{14}$

③ **a** 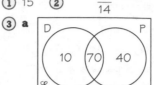   **b** $\frac{7}{11}$

Skills boost 1 Combining events

Guided practice

   a Number of combinations = 4 × 3 = 12

   b Number of combinations = 6 × 4 = 24

① **a** 20   **b** 21   **c** 15

② 18   ③ 36   ④ 25

⑤ 35   ⑥ 63   ⑦ *mn*

Skills boost 2 Tree diagrams

Guided practice

1st match   2nd match

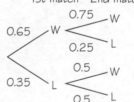

P(exactly one win) = P(W, L) + P(L, W)

= 0.65 × 0.25 + 0.35 × 0.5 = 0.3375

① $\frac{4}{7}$

② **a**

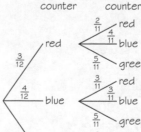

   b 0.91

③ 0.6475

④ **a** independent   **b** dependent
   c dependent   **d** independent

⑤ **a**   1st   2nd   **b** $\frac{47}{66}$
                counter   counter

⑥ $\frac{17}{35}$

Skills boost 3 Venn diagrams for conditional probability

Guided practice

a 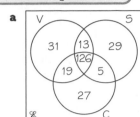 **b** $\frac{145}{189}$

① **a** 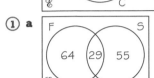 **b** $\frac{21}{37}$ **c** $\frac{29}{84}$

② **a** 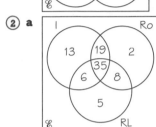 **b** $\frac{8}{11}$ **c** $\frac{27}{32}$

③ **a** 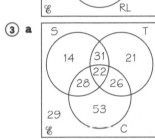 **b** $\frac{50}{129}$

④ **a** 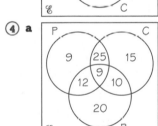 **b** $\frac{1}{4}$ **c** $\frac{19}{51}$

Practise the methods

Check up $\frac{4}{9}$

① **a** $\frac{1}{4}$ **b** $\frac{1}{2}$

② **a** $\frac{5}{9}$ **b** $\frac{2}{3}$

③ 21

④ $\frac{7}{15}$

⑤ **a** 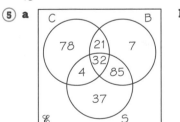 **b** $\frac{117}{158}$

Problem-solve!

① $\frac{127}{190}$

② **a** $\frac{7}{15}$ **b** $\frac{14}{15}$

---

③ $\frac{5}{14}$ **④** $\frac{7}{15}$ **⑤** 4

⑥ $\frac{10x - x^2}{45}$

⑦ **a** [Venn diagram: G, F, S with values 57, 73, 23, 31, 19, 14, 33] **b** $\frac{38}{125}$ **c** $\frac{104}{140} = \frac{26}{35}$

Unit 6 Direct and inverse proportion

AO1 Fluency check

① **a** **i** 50 **ii** ± 8
   b **i** 2.5 **ii** ± 0.4

② **a** $x = \pm 2\sqrt{y}$ **b** $x = \pm\sqrt{\dfrac{y}{3}}$

   c $x = \pm\sqrt{\dfrac{5}{y}}$ **d** $x = \pm\sqrt{\dfrac{b}{ay}}$

③ Number sense

8, the others are all square numbers

Confidence questions

① $x = \pm 20$

② $y = 0.0004$ or $y = \dfrac{4}{10\,000}$

③ No, a graph in which y is directly proportional to x is a straight line that passes through the origin.
This graph is not a straight line. (It shows a graph in which y is directly proportional to x^2.)

Skills boost 1 Direct proportion involving squares

Guided practice

a $y \propto x^2$
So $y = kx^2$
$75 = k \times 5^2$
$k = 3$
The formula is $y = 3x^2$

b When $x = 8$, $y = 192$

① **a** $y = 5x^2$ **b** $y = 180$

② **a** $y = 4x^2$ **b** $y = 196$ **c** $x = \pm 5$

③ **a** $q = \dfrac{1}{5}p^2$ **b** $q = 3.2$ **c** $p = \pm 0.5$

④ **a** $b = \dfrac{1}{2}a^2$ **b** $b = 72$ **c** $a = \pm 1.4$

⑤ $y = 18.49$

Skills boost 2 Inverse proportion involving squares

Guided practice

a $y \propto \dfrac{1}{x^2}$

So $y = \dfrac{k}{x^2}$

$0.04 = \dfrac{k}{10^2}$

$k = 4$

The formula is $y = \dfrac{4}{x^2}$

b When $x = 5$, $y = 0.16$ or $\dfrac{4}{25}$

① **a** $y = \dfrac{5}{x^2}$ **b** $y = 1.25$

② **a** $y = \dfrac{1}{2x^2}$ **b** $y = 3.125$ **c** $x = \pm0.02$

③ **a** $q = \dfrac{4}{5p^2}$ **b** $q = 0.032$ or $q = \dfrac{4}{125}$

  c $p = \pm0.4$

④ $y = 25$

Skills boost 3 Direct and inverse proportion graphs

Guided practice

In graphs A and D y is directly proportional to x.

① $a = 4$, $b = 17.5$
② $p = 2.5$, $q = 10$
③

Proportionality statement	Graph
y is directly proportional to x.	B
y is inversely proportional to x.	D
y is directly proportional to the square of x.	A
y is inversely proportional to the square of x.	C

Practise the methods

Check up $k = 10$

① A and iv, B and i, C and ii, D and iii
② $y = 8x^2$
③ $y = \dfrac{10}{x^2}$
④ **a** $y = \dfrac{5}{4}x^2$ **b** $y = 405$
⑤ graph B
⑥ $T = 25\,920$

Problem-solve!

① When n is doubled, D is multiplied by 4 as it is proportional to n^2.
② $b \propto a^3$
③ $y = 368\,640$
④ $y = 5.4$
⑤ $d = 1200$
⑥ $a = 0.5$, $b = 1.25$

Unit 7 Accuracy and bounds

A01 Fluency check

① **a** 4 **b** 5.84 **c** 10
  d 23\,000 **e** 24.7 **f** 5.6
② **a i** 12.5 cm **ii** 11.5 cm
  b i 5.235 km **ii** 5.225 km
  c i 7500 kg **ii** 6500 kg
  d i 3.65 litres **ii** 3.55 litres
  e i 8550 m **ii** 8450 m
  f i 6.45 km **ii** 6.35 km

③ **Number sense**

  a 1.35 **b** 4.25 **c** 10 **d** 38

Confidence questions

① 33.7125 mm² and 32.5325 mm²
② 7.935 g/cm³ (see guidance in callout p49) and 7.804 g/cm³

Skills boost 1 Upper and lower bounds of measurements

Guided practice

Width = 7.9 cm (to 1 d.p.)
Upper bound = 7.95 cm, lower bound = 7.85 cm
Length = 12.4 cm (to 1 d.p.)
Upper bound = 12.45 cm, lower bound = 12.35 cm
Area = length × width
Upper bound = 7.95 × 12.45 = 98.9775 cm²
Lower bound = 7.85 × 12.35 = 96.9475 cm²

① 11.1375 cm² and 10.4675 cm²
② **a** 21.6 cm ≤ P < 22.0 cm
  b 21.4 cm ≤ P < 21.8 cm
③ 6.45π cm ≤ C < 6.55π cm
④ **a** 21.2175 cm² ≤ A < 22.1875 cm²
  b 26.07 cm² ≤ A < 27.135 cm²
⑤ 4.77 cm

Skills boost 2 Upper and lower bounds of compound measures

Guided practice

Distance = 100 m (to the nearest metre)
Upper bound = 100.5 m, lower bound = 99.5 m
Time = 11.2 s (to 1 d.p.)
Upper bound = 11.25 s, lower bound = 11.15 s
Speed = $\dfrac{\text{distance}}{\text{time}}$

Upper bound for speed = $\dfrac{\text{distance upper bound}}{\text{time lower bound}}$
  = 100.5 ÷ 11.15
  = 9.013 m/s (3 s.f.)

Lower bound for speed = $\dfrac{\text{distance lower bound}}{\text{time upper bound}}$
  = 99.5 ÷ 11.25
  = 8.84 m/s (3 s.f.)

① 3.45 hours = 3 hours and 26.9 minutes and 3.42 hours = 3 hours and 25.3 minutes
② 20.53 g/cm³ and 18.26 g/cm³
③ 18.89 N/m² and 17.66 N/m²
④ 7.28 g/cm³ (3 s.f.)

Practise the methods

Check up 7.944 g/cm³

① **a** 5.17 and 4.19
  b 21.63 and 20.65
  c 48.57 and 40.54
② **a** 6.81 and 6.52
  b 93.80 and 90.65
  c 60.11 and 48.65
③ 20.29 N/m² and 18.65 N/m²
④ 6.149 cm
⑤ 1.06 g/cm³

Problem-solve!

① 0.061 (3 s.f.)
② 47.6 mph (2 d.p.)
③ 1.22 g/cm³ (3 s.f.)
④ 2.24 seconds (3 s.f.)
⑤ 7497 m²